# HOW TO GET ON HF THE EASY WAY

## TABLE OF CONTENTS

# HOW TO GET ON HF THE EASY WAY

By: Craig E. "Buck," K4IA

ABOUT THE AUTHOR: "Buck," as known on the air, was first licensed in the mid-sixties as a young teenager. Today, he holds an Amateur Extra Class Radio License.

Buck stands on the ARRL Honor Roll with 337 countries out of 340 confirmed and has earned the 8 Band DXCC award. He is also an active instructor and a Volunteer Examiner. The Rappahannock Valley amateur radio Club named Buck the Elmer (Trainer) of the Year three times.

Email: K4ia@EasyWayhamBooks.com

Published by EasyWayhamBooks
130 Caroline St. Fredericksburg, Virginia 22401

**How to Get on HF – The Easy Way** and other Easy Way books by Craig Buck are available at Ham Radio Outlet and GigaParts stores and from Amazon:
**How to Chase, Work & Confirm DX**
**Pass Your Amateur Radio Technician Class Test**
**Pass Your Amateur Radio General Class Test**
**Pass Your Amateur Radio Extra Class Test**
**Prepper Communications**

ISBN 978-1535171861

5.0

# INTRODUCTION

I've taught many classes and found new hams are intimidated by HF (3–30 MHz). They fear the big leagues and stay on the local repeater until they get bored and burn out, missing an awaiting world.

Ham tests (even the Extra Class test) do not teach how to get on the air and operate. You earned your license, now what? How do you choose and assemble the needed equipment? How to make your first QSO? This book shows The Easy Way to get on HF.

I avoid endorsing any manufacturer or product. Occasionally, I might mention a product to describe its functionality. Read the reviews on eHam.net, *QST* magazine, and other sources online and in print for comparisons.

There are opinions on these pages, and there may be other viewpoints or qualifiers. Please accept generalizations for the sake of simplicity. Here's something I discovered a long time ago: "If you ask five hams a question, you will get seven different answers and maybe a fistfight."

I am not an electrical engineer; I flunked math and went to law school. I have been a ham radio operator long enough to offer valuable insight. My approach is practical, not technical. No apology for that.

Please send comments, corrections, and chastisements to me at K4ia@EasyWayhamBooks.com This book prints on-demand. Edits can appear within 24 hours.

# HOW I GOT STARTED

If you've read any of my other books[1], you have already heard a lot of my story. I always begin at the beginning so, please ride along for a while.

I am a product of the days before personal computers, the Internet, video games, and 500 channels of color TV. Dad was an Army officer, and I grew up in post-war Germany when there was no television. I didn't see a TV set until fifth grade. Now, I consider that deprivation to have been a blessing. It taught me to read and entertain myself instead of sitting sedated in front of the black box, mesmerized by a flickering wasteland.

We made our entertainment building an airplane cockpit using shoe boxes with toothpicks for control levers. There were model airplanes and ships. One year, Christmas brought a chemistry set, another an Erector set.

We played in the woods, caught snakes, lit cherry bombs, and experimented with home-made black powder and fireworks. A kid could go to jail for that today. In fifth-grade, we built a crystal radio set. It only tuned one station. I would tuck the earpiece under my pillow at night and fall asleep listening to basketball games. Basketball on the radio took a lot of imagination and did not interest me. Radio coming from a rock had my attention. The station was about 30 miles away.

A Heathkit shortwave receiver appeared under my seventh-grade Christmas tree. I was already a builder and looked forward to assembling and soldering. Miraculously, the radio worked.

---

[1] Listed on the front cover sheet

## HOW I GOT STARTED

Nothing prepared me for what came through that speaker. It was eerie hearing Radio Havana Cuba or Radio Moscow during the Cold War. I expected the FBI to break down the door. My first lesson in political "spin" came from propaganda. The tune to identify Radio Havana was "Look Sharp, Be Sharp," the same used in Gillette Razor Company commercials. I guessed since Castro had a beard, he didn't know about Gillette razors.

Airline pilot, shipbuilder, chemist, or architect wasn't in my future. However, that receiver was my introduction to the world. The Cold War was in full swing, and there was a battle for hearts and minds carried out on shortwave radio. Voice of America, Radio Free Europe, and the BBC competed against the likes of Radio Moscow and Radio Havana Cuba. Virtually every country had shortwave service, and most had some English-language programming. I didn't know or care what they were talking about, but it was exciting to hear a signal from another part of the world. My young imagination led me on travels around the globe.

Tuning across the bands, I heard people talking to each other. That was how I discovered amateur radio. I knew that was something I wanted to do. Listening to distant stations, fading in and out of the static, sometimes with distinctive polar echoes, carried a sense of magic. If I became a ham, I could graduate from passive listener to participant.

We were experimenters and builders—kids with curiosity. Today, we would be called "makers." Other than globe-trotting, what was the lure of ham radio? It was the first time all the stuff they were trying to teach me in school had a use. How many times have you heard adolescents complain, "I don't need to know algebra." "I'll never use chemistry." "Why do I have to learn about the ionosphere?" "Who cares about Pitcairn Island?" And on and on.

We use algebra to determine antenna lengths. Study of the ionosphere tells the best time to talk to Australia. Battery chemistry decides the best power source for your equipment. Geography will determine which direction to orient an antenna. Knowing a few words in another language lets you say "Hi" to a foreigner in his native tongue. Astronomy teaches meteor scatter, sunspots and when to bounce a signal off the moon or satellite. Physics can teach how to design an antenna. History yields an appreciation of Pitcairn Island. I could teach most of an entire high school curriculum based on amateur radio.

And what does Pitcairn Island have to do with it? Plenty. During the summer, Mom would take me to the library. I enjoyed the tall-ship swash-buckling tales with pirates and mutinies. My favorite was The Mutiny on the Bounty, a true story from 1789 with real people, a real mutiny and real villains and heroes. You decide who the villains were and who were the heroes. The mutineers, led by Fletcher Christian, fled to Pitcairn Island in a remote part of the Pacific Ocean.

After I got my license, I was tuning the bands late one night, and I heard VR6TC, Tom Christian, Fletcher's descendant living on Pitcairn. VR6TC was Tom's callsign. I "worked" him, and I was shaking. Tom was famous because of the Bounty story and the rarity of his location. Tom and his wife got their amateur radio licenses because remote Pitcairn had no other means to communicate with the outside world. I never heard Tom on the air again, and he became a Silent Key[2] a few years ago.

---

[2] A deceased ham is referred to as a "Silent Key," a throw-back to the telegraph era. You will see it as VR6TC (SK)

## HOW I GOT STARTED

English class required a "What I did Last Summer" essay. I wrote about contacting Tom Christian. My teacher didn't believe me. She said the assignment wasn't supposed to be fiction. Months later, I received a QSL card[3] from Tom. I enjoyed rubbing her nose in it. Maybe that's why she gave me a C in English that quarter. Gloating is never wise.

K4IA as WA4TUF in 1967

I went QRT[4] after college and lost Tom's QSL card, but you can see it on the wall to the top left edge of the map in my picture. I have worked and confirmed Pitcairn many times since, but nothing can replace my Tom Christian card. My contact with Tom is one I will never forget.

That picture with a microphone is deceptive. The microphone was hardly used. My squeaky voice was a

---

[3] Written confirmation of our contact.
[4] QRT means "Off the air."

dead giveaway that I was "just a kid," so I became proficient on Morse code, and no one knew my secret.

There was very little activity on the VHF/UHF bands before repeaters came into use in the 1970s. The Technician License was not as popular as it is today. Most of us went straight to HF operating.

I borrowed my first rig, a very simple one-tube 40-meter homebrew transmitter. Transceivers (transmitter and receiver in one box) were for rich kids, so I had a separate receiver, also borrowed. The first antenna was a 40-meter dipole, sixty-six feet of wire. Nothing fancy, but I had a blast.

I bring up my modest beginning to encourage you. It doesn't take thousands of dollars to get on the air. You do not need to buy the latest and greatest WizBang 4000 Pro. Lots of stations, including DX[5], will respond to simple equipment. The most important piece of advice I can give is, "Get on the air with whatever you can cobble together."

We will get into equipment selection later.

A present-day Technician has limited HF privileges and a maximum power of 200 watts PEP[6] output. The Technician Class HF frequencies are:

3.525-3.600 MHz: CW only
7.025-7.125 MHz: CW only
21.025-21.200 MHz: CW only
28.000-28.300 MHz: CW, RTTY/Data
28.300-28.500 MHz: CW, Phone

---

[5] DX is "distance" and used to designate stations outside the continental US.
[6] PEP is "peak envelope power."

## HOW I GOT STARTED

Most Technician HF privileges are CW, but the data and phone allocations on 10 meters (28 MHz) can produce some amazing contacts when the band is open.  Why limit yourself to one band and sporadic openings? To be effective on HF, get your General.

The General Class license will give you HF privileges on parts of all the bands.  Go for it.  The test is not hard, and my book, "Pass Your Amateur Radio General Class Test – The Easy Way," will get you through.

Dig in and discover the magic of HF.

Mount Athos, Greece is an ancient, cloistered monastery. Monk Apollo (SK), was its only contact with the outside world.

# GETTING STARTED

HF can be intimidating.  Here are some essential tips to get started:

- Join your local radio club.
- Find an Elmer.
- Attend hamfests.
- Take part in Field Day.
- Get on the air.

**Join Your Local Radio Club**  The local club will allow you to rub elbows with experienced HF operators and learn from them.

Lists of local radio organizations are at ARRL.org/find-a-club.  There are clubs everywhere.  Most host a repeater.  Listen in and get on the repeater as your introduction to the club.  Then, attend a meeting.  Most clubs are eager to receive new members and encourage new hams.  Don't be intimidated if the discussion is over your head; you just need time to catch up.

Meet some folks.  Take part in club activities.  See if the group is a fit for you.  Some clubs emphasize public service, providing communications for marathons, bike races, and other events.  Other clubs may be interested in contesting or just eating breakfast.  If you don't feel the love, try another organization.

**Find An Elmer**  Joining a local club is a great way to fulfill the second piece of advice: find an Elmer.  An Elmer is a mentor/coach.  This isn't a formal commitment.  You're not "going steady" or anything like that.  Look for someone who seems friendly, more knowledgeable, and willing to answer questions patiently. Your Elmer is your "go-to" guy when you get stumped or don't understand the jargon.

You might find more than one Elmer. One is the guru of antennas; another is a contesting fanatic, and another likes computers. You will find most hams willing to share their expertise, so don't be afraid to ask.

**Attend hamfests** A hamfest is a gathering to demo, swap, buy, or sell ham equipment. Many include forums and food where you'll meet for an eyeball QSO[7] and enjoy a good time. It is a convention that mixes a flea market, museum, vendors, and camaraderie. You will often find reasonably priced used equipment along with connectors and cables for less than the cost to ship if ordered online.

Hamfests can be elaborate affairs attended by tens of thousands such as the annual Dayton Hamvention in May, the Visalia, California DX Convention in April, or the Orlando Hamcation in February. There are also numerous local hamfests and low-key tailgate "junk-in-the-trunk" gatherings. You can find lists of upcoming events at arrl.org/hamfests-and-conventions-calendar.

**Participate In Field Day** Field Day is a long-standing amateur radio tradition. The fourth full weekend in June hams take to the great outdoors and operate using emergency power.[8] The goal is to set up a station and contact as many other stations as possible during the 24-hour activity. This event is a combination of emergency-preparedness training, contest, picnic, and camp out.

---

[7] Meet face-to-face.

[8] There are other operating categories as well, but this is the main one.

Your local club probably takes part in Field Day. Here's an opportunity to see antennas erected, stations set up and operators operating. You can take part as a helper or operator.

The rules allow for a Get-On-The-Air (GOTA) station restricted to new or inactive hams. An unlicensed person may operate under supervision. Ask about the GOTA station and take advantage of the opportunity. Don't miss Field Day. It is an exciting time.

**Get On The Air** The last piece of advice is, "Get on the air." The way to get started is to start. No matter how modest your station is, you can make contacts. Operating tips and techniques are coming up in another chapter.

If you don't have a station set up yet, see if a local will let you come over and operate, or listen to him operate. Sit in on a contesting station for some serious radiosport excitement.

Call CQ. I use QSL cards from my collection to illustrate the points in this book. I hope you enjoy them.

# MODES

The mode is the transmission type. You have probably operated FM mode on VHF. HF opens a whole new world of possibilities. Before we talk about equipment and outfitting your station, let's review the various modes of operation.

This German ham, operating in Senegal, West Africa must have had a hard time picking a favorite mode.

## SINGLE SIDEBAND (SSB)

SSB is a voice mode, and the most significant impediment for beginning hams on SSB is "mic fright." If you operate on the local repeater and have never ventured into the unknown world beyond your usual gang of local friends, you might be reluctant. Get over it. You got over mic fright after the first couple of times on the repeater and can do it again. The guy on the other end might be as nervous as you. Assume he is and practice on him.

Don't let DX stations intimidate you. An experienced DXer has heard it all and will help you through the contact. Get over your jitters by working Europeans. They are plentiful and friendly.

The difficulties with SSB are:

- It is a wide mode (3 kHz) requiring extra bandwidth. Only a few stations can fit in the listener's spectrum, so there is more interference (QRM).
- Voice communication is difficult to understand through interference.
- SSB takes more power than CW to beat the signal to noise ratio.
- The spoken word is easy to misunderstand.

When operating SSB, pay attention to audio quality and equipment adjustments. The default adjustments on your transceiver may be just fine for casual operating. Start there and don't fiddle just for the sake of it. "If it ain't broke, fix it 'til it is broke." Enlist a friend to make critical comments on the signal.

There are steps to improve your audio. First, the audio should fill the signal. This is hard to describe, but you can sense when the audio should be louder. It is obvious on the repeater with weak audio but a strong full-quieting carrier. Usually, the operator is speaking too far from the microphone or facing away from it. Talk close to the microphone and in a normal tone of voice. If you are too far away and try to compensate by increasing the mic gain, you pick up more background noise, such as fans, air conditioners, or televisions.

Adjust your mic gain for full but proper drive. Over-driving, which can cause splatter and distortion, is worse than under-driving. Refer to your equipment manual for details on adjusting your transmitter.

## MODES

Once you have the audio drive properly adjusted, you can tailor your audio response frequency. Adjust the speech processor emphasizing readability, not fidelity. Speech is best understood between 2,000-3,000 Hz. Lower frequencies use more power but don't improve clarity. Roll off the lower frequencies and increase the upper-middle frequencies in a speech processor or by your selection of a microphone. That will not sound broadcast quality but will concentrate power in the spectrum essential for understanding. After adjusting the range, go back and tweak the gain again.

Compression is another adjustment. It evens out the audio level, providing some extra "punch" to quieter syllables. It doesn't compress the loud but boosts the quiet. Don't use too much or the signal will "ring," distort or pick up background noises.

## SSB OPERATING

HF phone differs from the local repeater. Casual repeater use is informal chit-chat conducted over clear connections. You operate on a memory channel and don't spin a VFO dial. Never call CQ on a repeater; instead. say your callsign, "K4IA listening."

HF presents some new challenges. First, be aware of the license class band edges. You might program a favorite net into a memory channel, but most HF operation is VFO controlled, and it is easy to slide where you are not allowed. Don't rely on the few frequencies memorized for the license test. Consult a chart to be safe. Most hams have a frequency chart at their desk.

Second, listening is much more challenging on HF. It requires tuning around to find contacts. Be aware of nearby stations. Interference is rare on a repeater, but it is very common on HF. Listen and ask if the frequency is in use before you call CQ. Listen above and below your frequency.

Stay away from the band edges and other conversations. Your signal is wider than you think. On SSB, the signal can extend 3 kHz above (USB) or below (LSB) the dial frequency. If the dial frequency is close to the band edge, the signal is over the edge.

The recommended separation on SSB is 3kHz, but stations often operate much closer. It is easy to bleed over into someone else's contact, particularly when propagation changes. A clear frequency can get crowded quickly. If it does, and someone complains, they might not be pleasant. Rise above it. If someone interferes with your conversation, be polite. Go to their frequency and ask them nicely to move a bit. Or you can move your conversation to a clearer spot. No one owns a frequency, but arguing will ruin everyone's enjoyment.

There are more tips to help you with SSB operating in the chapter "Anatomy of a QSO – The Right Way."

## CW (MORSE CODE)

"Learning Morse code" is a later chapter. Here is what to know about the mode.

CW, which stands for continuous wave or carrier wave, is the most basic of digital modes. The carrier is turned on and off, making the dots and dashes read as Morse code.

Send Morse with a straight key, semi-automatic key, electronic keyer, or computer software. A key pumps up and down to send individual characters. A semi-automatic mechanical key is sometimes called a "bug." Push in one direction, and a swinging arm sends a

string of dits.  Push the other to send individual dahs.
Only the dits are automatic.

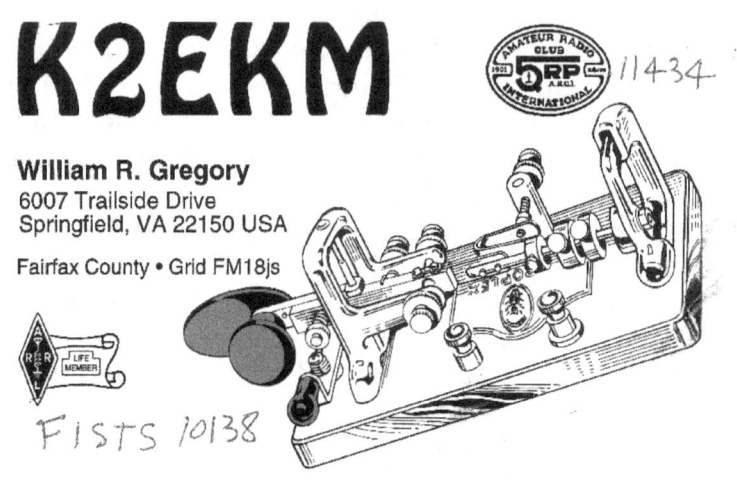

**K2EKM**  11434

**William R. Gregory**
6007 Trailside Drive
Springfield, VA 22150 USA

Fairfax County • Grid FM18js

FISTS 10138

This is a "bug" keyer.

Paddles use an electronic keyer, perhaps in your radio,
to send characters automatically.  One direction sends
continuous dots and the other successive dashes.

Most CW operators decode CW manually, either
writing it down or in their heads.  A computer program
or stand-alone code reader will work reasonably well
with clear signals and perfectly sent computer-
generated characters.  I would not encourage you to
use such a reader, as it will become a crutch that
discourages learning.  The human ear and brain are
best.

A CW signal is readable 10-13 dB below the level of an
SSB signal.  That is a considerable advantage for CW.
A good CW operator can copy at the noise level or
slightly below, and that is just about impossible with
SSB.  Since every 10 dB represents ten times the
power, a 100-watt CW signal can be the equivalent of
a 1,000 watt SSB signal.  That is plenty of incentive to
learn CW!

My good friend Pietro Begali makes unique CW keys and paddles. This model is the Sculpture, one of my favorites.

## CW OPERATING

CW signals are narrower than SSB, with a transmitted signal approximately 150 Hz wide. That means 20 CW signals can fit into the same space as one SSB signal.

The recommended separation for CW is broader, 150 – 500 Hz. A typical filter used for CW is 500 Hz wide, although many receivers can go narrower. Narrower filters cut down interference and noise outside their passband.

Use a broad filter while searching for a signal. You will hear more and not tune past a station. Switch to a narrow filter once you find something. Copying is all about the signal-to-noise ratio (SNR) and narrowing the passband will reduce noise and improve copy.

Morse code frequently uses abbreviations and Q signals. A few Q signals were on the General test, and there are about a dozen in everyday use. The Q signal embeds a long message into three letters. Add a "?" to ask a question. QTH means "location." "QTH?" means, "What is your location?" Another advantage of Q signals is they say the same thing in different languages. If I tell a Russian to "QRS," he knows I mean to slow down even though neither of us speaks the other's language. There is a list of common Q signals in the Appendix.

Morse code also uses abbreviations to speed up the conversation. For instance, "tt" means "that" and "tmrw" is tomorrow. "De" means "this is." Internationally recognized abbreviations make Morse a universal language. The Appendix has a list of standard abbreviations.

There are Morse code characters for punctuation marks, but the only one commonly used is the question mark. We don't use or need commas, periods, dollar signs, and parentheses.

Prosigns are a combination of letters, sent as one. The most common are:
BK – break, as in "over"
KN – don't break in on our contact
SK – end of contact
DN - The slash "/" character (DN sent as one letter) added to callsigns identifies, mobile, K4IA/M, or portable, K4IA/P, operations. I was 4X4/K4IA in Israel.

The recommended procedures for SSB in the "Anatomy of a QSO" chapter hold for CW as well. Short CQs, short acknowledgments and don't repeat data unless asked.

Here's an example of how <u>not</u> to do it:
"WA4TUF WA4TUF DE K4IA K4IA K4IA  THANKS FOR THE CALL.  YOUR RST RST IS 579 579 579.  MY QTH QTH IS FREDERICKSBURG FREDERICKSBURG, VA VA.  NAME NAME IS BUCK BUCK BUCK.  SO HOW DO YOU COPY ME? WA4TUF DE K4IA K"

My typical exchange might go like this:
"WA4TUF DE K4IA  GM OM  UR 579 FREDERICKSBURG VA  OP BUCK BUCK  HW?  BK"

No repetition and no punctuation, just a slightly longer pause after a sentence.  If I feel lazy, I leave out "Fredericksburg."  I repeat my name because it is unusual.  If your name is "Bob" you probably don't need to repeat it.

End the conversation with the "final."
"TNX QSO CRAIG  HPE CU AGN SN  73 ES BEST DX WA4TUF DE K4IA  SK"

The final return is often followed by a dit-dit to which the other station replies with a single dit as the last acknowledgment.

## DIGITAL MODES

There are many digital modes, but I limit this discussion to PSK, RTTY and FT8 as they are the most common.  Digital modes are a good way to start on HF.  You can make many contacts with low power and a minimal antenna system while avoiding mic fright, and you don't need to know Morse code.

PSK caught on quickly after introduced in 1998.  PSK relies on the computer soundcard to generate a warbling audio signal that is fed to the transceiver for transmission.  To receive the audio output from the transceiver feeds to the soundcard.  Computer software converts the received signal to letters, and

## MODES

they trot across the screen about as fast as you can type.

PSK has several versions, with PSK31 being the most popular. A PSK31 signal is only 60 Hz wide, much narrower than CW (150 Hz). Because PSK is narrow and the decoding software robust, PSK can operate at very low signal levels. PSK can decode a signal seen on a waterfall display even though it is too weak to hear. Typically, output power is limited to 25 watts or less. PSK is ideal for low power operations.

There are many free PSK programs available, and they do not require a super-computer to operate. Examples are WinWarbler, Digipan, and Fldigi. Some work with a tablet or smartphone. The audio hookup from the computer to the radio does not have to be complicated, although it is advisable to have some isolation to prevent ground loops and RF feedback. Check out the reviews for "Interfaces" on eham.net or roll your own.

Winwarbler is part of the free DxLab Suite of software by AA6YQ, Dave. I have used DxLab for years and, without meaning to disparage any of the other programs, heartily endorse it. There's more about logging and record keeping software in another chapter.

With narrow bandwidth, robust decoding, and low power, it is easy to see why PSK is popular. You will see many stations operating just about any time you look. You will also find many DX stations.

PSK is a conversational mode, unlike FT8 which only transmits location and signal report. More on that later.

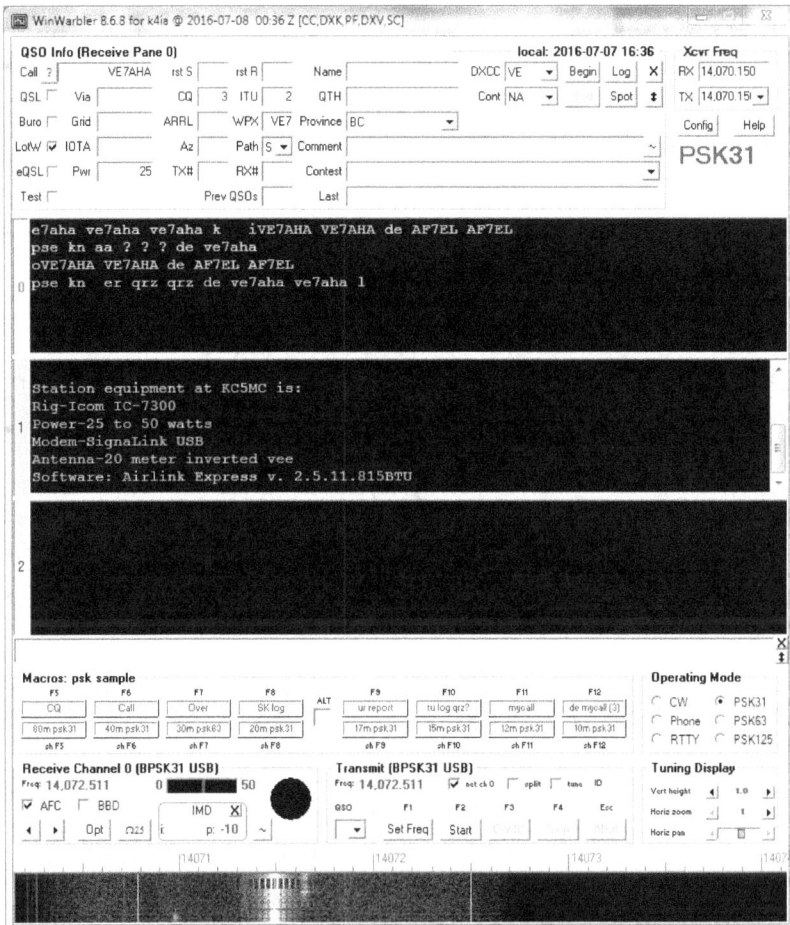

Here is a screenshot from WinWarbler in PSK mode.

The center three windows show different decoded PSK conversations. The waterfall display at the bottom shows signals in the passband. With your computer connected to the radio, clicking on a waterfall signal will change the radio to that frequency. Clicking on a call sign enters it and the frequency information in the log portion at the top. Macro buttons allow you to store pre-programmed messages so you don't have to be an accomplished typist. For instance, you can program a button with your name, QTH, and equipment.

## MODES

RTTY, or teletype, formerly used mechanical printers communicating over a wire. Modern RTTY is another soundcard mode that relies on computer power for encoding and decoding.

RTTY is faster than PSK and occupies a much wider bandwidth. RTTY operators can run full power, but RTTY uses a 100% duty cycle, so your equipment has to be stout. A few minutes of off-and-on calling should be safe. If the transceiver cooling fans come on too often, or the case feels hot, try reducing power a bit.

Winwarbler, described above, will operate in RTTY mode as will many other free programs. You will almost always see RTTY signals on the bands, and there are several very active RTTY contests during the year.

## PSK and RTTY OPERATING

These digital modes make extensive use of pre-recorded Macros (canned messages). Pressing a Function key button on the computer keyboard sends the message. Some messages may embed data from the log entry (QSO #), the operator's name, or weather reports gleaned from the Internet.

PSK and RTTY have different personalities. PSK Macros tend to be long and chatty RTTY is terser.

PSK coding uses varicode; the length of individual letters is different. Capital letters take twice as long to send, so avoid using all caps with PSK. RTTY, on the other hand, is always in capital letters. There isn't a choice.

You will notice longer CQs on the digital modes. That is because it takes a little longer for someone to see the signal, click on it, and start decoding. You often see a final "CQ" appended to the end of the call to catch a listener who tuned in late. Here's an example of a macro that puts the rig into transmit mode, sends CQ, then returns to receive mode. We'll assign CQ to the F1 key.

F1 <start> CQ CQ CQ de K4IA K4IA CQ CQ CQ de K4IA K4IA CQ <stop>

Most digital programs come with a set of canned macros ready to customize. Learn more about digital mode operating by watching conversations and developing your own macros.

## FT8 OPERATING

FT8 and its cousin, FT4, have become wildly popular accounting for the majority number of ham contacts. The free software is WSJT[9] and its derivatives.

One reason for its popularity is that FT8 will decode signals too weak to hear. Another is that the computer automates the process by doing most of the work. Each side of a conversation takes 15 seconds exchanging a signal report and location. The complete QSO is over in less than 2 minutes.

FT8 is not a conversational mode. It is impersonal, but will fill a logbook with worldwide contacts.

---

[9] Weak Signal Joe Taylor, named after the inventor.

# ASSEMBLING A STATION

I refrain from recommending particular models or manufacturers. We all have our favorites, and some people like Fords while others favor Chevys. As my beloved Nana used to say, "That's why they don't make all ice-cream vanilla." Here are some criteria to consider when choosing equipment.

## HOW MUCH DO YOU WANT TO SPEND?

Amateur radio isn't cheap, but the cost of a video gaming system or smart phone gets you on the air. The initial expenditure may seem high, but amateur radio is cheap compared to SCUBA diving, golf, or a bass boat.

Take comfort in the fact amateur gear holds value and is easy to re-sell. Consider the annual cost. If you choose to go all-out with new equipment, in five years, a $2,000 transceiver might fetch 75% of the original price. That would be a discount of $500 for five years of enjoyment or $100 a year. That amount wouldn't buy much golfing.

Used equipment depreciates even slower. Pay $500 for a used transceiver, and you might reasonably expect to sell it for very close to that in a few years. Say it sold for $450 after two years; the cost was only $25 a year.

Some hams never sell the old gear and accumulate a museum in their closet. I often sell the last rig when I upgrade to reduce the incremental cost of upgrading. Then, nostalgia kicks in, and I regret the sale.

Still, the sticker shock of the first purchase can be formidable. Decide how much you want to spend and leave room for some extras. Here are a couple of sample budgets.

**Minimal**  The most minimal setup is low power, possibly kit form, and Morse code or digital modes on one frequency.  I only recommend these to more experienced operators.  A beginner would be frustrated easily.

**Minimal Station**

| QRP[10] Transceiver | $50 - $250 |
|---|---|
| Single Band Wire Antenna | $25 |
| Feed line | $25 - $50 |
| Power Supply (batteries) | $25 |
| Misc[11] | $100 |
| TOTAL | $200 - $450 |

**Moderate Used**  A moderate used station would be an all-mode transceiver five to fifteen years old.  It will cover 80 through 10 meters.  This equipment is modern enough to give plenty of service, but won't be cutting edge technology.  It might not have been top-of-the-line even in its day.  The higher end of the price range would give you something newer or a more fully equipped older rig.  You can expect computer control, good filtering, all-solid-state, and an antenna tuner but not the latest-and-greatest of these features.

**Moderate Used Station**

| Used HF Transceiver | $500- $1,000 |
|---|---|
| Used Antenna Tuner | $100 |
| Multi-Band Wire Antenna | $75 - $150 |
| Feed line | $50 - $100 |
| Used Power Supply | $100 |
| Misc | $100 |
| TOTAL | $925 - $1,550 |

**Moderate New Station**  The following chart describes an all-new station with average level

---

[10] QRP means low power, generally less than 5 watts.
[11] CW key, antenna rope, insulators, battery charger

components. This isn't the most expensive gear available, but it also isn't the cheapest.

Many newer radios are based on software rather than discrete components. As a result, moderately priced new radios can deliver better results for less money than older models. Today's $1,000 entry-level transceiver has specs that equal or exceed older rigs costing much more. The latest radios have built-in panadapters displaying signals across the band. My overall recommendation is a moderate new station.

## Moderate New Station:

| Mid-range HF Transceiver | $1,000 - $2,000 |
|---|---|
| New Antenna Tuner[12] | $150 - $250 |
| Multi-Band Wire Antenna | $75 - $150 |
| Feedline | $50 - $100 |
| Power Supply | $125 |
| Misc | $100 |
| TOTAL | $1,500 - $2,725 |

There is a lot of variance. These figures are very rough estimates designed. Anything you can beg or borrow will reduce the initial cost. We'll talk about the individual components in later chapters.

I did not include towers, beam antennas, high-end radios, and other sophisticated equipment because they are beyond getting started on HF. I doubt you will want to spend $13,000 on your first transceiver. Start simple and grow into the more sophisticated and expensive offerings.

---

[12] Maybe, you won't need one if your transceiver has an internal tuner. See more in the chapter on antennas.

# USED VS NEW

Much like computers, radio equipment has undergone incredible technological advances. However, radio itself hasn't changed, and there is a lot of 50 to 60-year-old gear still in service. You can save some money with used equipment, but what are the considerations when buying used?

Vacuum tube gear is a collector's item. Some of it still works just fine. I have a Drake transceiver I enjoy very much. It is almost 60-years old. However, parts and tubes can be hard to find, and tube radios are projects much like an antique car. I would not recommend you start with vacuum tube gear unless someone gives it to you.

Finding replacement parts for old solid-state gear can also be a problem. By "old" I will arbitrarily say 20 years. It's not that you can't find a particular resistor. It is the no-longer-made band switches, displays and mechanical pieces that are rare as unobtainium. Older radios can lose alignment, and parts may change value with time. Keeping old equipment operating is a hobby unto itself. Better stay away unless you are comfortable troubleshooting and fabricating.

Having warned you off older gear[13], I must add a qualifier. There are a lot of vintage solid-state and hybrid (solid-state except for tubes in the final amplifier) transceivers still performing excellent service. Take one as a loaner or a gift but don't pay a lot, $250 - $350 maximum. Make sure it works properly because the cost of a professional repair (with shipping back and forth) might be more than the radio is worth.

---

[13] Sometimes called "boat anchors."

## ASSEMBLING A STATION

Ten to fifteen-year-old gear is priced temptingly at half the new cost.  Future depreciation will be slow.  It will sell for close to what you paid as long as nothing breaks.  Check the reviews online (eHam.net).  See if a particular issue plagues the radio, such as failing displays, bad final transistors, or unobtainable replacement parts.

Five-year-old radios are practically new and sell for 75-80% of their original price if they are still in production.  These can be a value if the radio works properly.

There was a time when you could trust a ham not to sell a problem radio without full disclosure.  I don't know if you can count on that courtesy any longer.  Know your seller.  Buying from a local ham or reputable dealer at a hamfest is probably safe.  Don't expect a guarantee.  If you want a warranty, buy something new.

Buying online is riskier, though I have experienced good luck with sellers who had high positive feedback.  EBay and PayPal have strong buyer protection policies, so you have recourse if the radio is not as represented, dead, or damaged on arrival.  They can't help if it blows up after a week, even if you suspect the seller knew something was going bad.

Other online sites don't have the same buyer protections, so understand the terms before you buy.  Be careful of online scams.  If the price is too good to be true, it is.  If the seller says he can't test the gear, it is probably not working.

Try to borrow gear.  The local club may loan equipment to new hams.  A member might have a radio sitting in the bottom of his closet and be willing to let you use it.  Finding equipment is another good reason to join your local club.

If you borrow, be a good steward and treat the gear carefully. Have someone show you how to use the radio or spend time in the instruction manual before doing any damage. A missing manual is often available online.

Get a clear understanding of liability. Is the responsibility, "If it breaks in your possession, you bought it" or "I understand it is old and feeble, so whatever happens, happens?"

Agree, in advance, on the terms of your loan. I once let out a radio with no clear understanding, and it took me three years to get it back. If I loan again, I will take a picture of the borrower with the equipment and a sign that says, "Return by Jan 1, 2024." That gets the point across and helps everyone remember. On January 1, I can send him the picture as a hint.

Vintage Equipment at this German station.

# CHOOSING EQUIPMENT

Let's talk about the individual components that go into a station.  What features are best for the beginner on a budget?  Contesters and DXers will have more exacting expectations, but here is where we start.

## POWER SUPPLY

The power supply is certainly not the most exciting piece of equipment in the ham shack, but it is the heart.  Most modern radios do not have a built-in supply, and you'll need to provide one.  Here are the considerations that go into your decision.

Solid-state transceivers run on a nominal 13.8 volts DC, and you must convert your house voltage (nominal 120 volts, AC) to that level.  At 100 watts output, your transceiver will probably draw 1½ to 2 amps from the 120-volt outlet That converts to around 17 amps of current on the 13.8-volt side.  Add a few amps for overhead such as the other circuits in the radio, dial lights, etc. and you can estimate your supply should provide at least 20 amps.

The amount of current a power supply can provide before the voltage sags or the fuse blows is the supply's rating.  Don't scrimp and get a 20-amp power supply for a 100-watt transceiver.  It will be marginal and leave you no room for any accessories, such as a VHF radio in the shack.  The additional cost of a 25-amp or better, a 35-amp power supply is not significant, and you won't be sorry.  I like volt and amp meters on mine so I can see what is happening.  Meters don't add much to the cost, either.

There are two designs for power supplies.  The linear types use a heavy transformer.  They are very rugged.  Mine has been on almost continuously for 20 years without a glitch.

Switching power supplies convert the 60-cycle house AC to a higher frequency allowing for much smaller and lighter transformers. The conversion can generate RFI,[14] and switcher supplies have gotten a bad rap on that account. Check the reviews and pay particular attention to complaints about RF hash. Consider a switcher if you need to move the power supply often or carry it traveling.

Keep power supplies away from the transceiver to prevent coupling noise or hum into the radio and cables. My power supply is on the floor under the desk.

A new transceiver will come with a sufficiently stout power supply cable and the proper plug to fit the radio. A used transceiver may have neither. Don't scrimp with cheap speaker wire to power your rig. Resistance in the wire will drop the voltage. (Remember Ohm's Law?)

If your transceiver draws twenty amps, twelve feet of number 16 wire will drop the supply voltage by two volts at the radio.[15] Low voltage starves the transmitter of the power it needs and will distort your signal or shut down the rig.

Twelve feet of number 10 wire will only drop about half a volt. Use 8 or 10 gauge wire and make it short. If you use number 12, keep the length to six feet or shorter. The proper supply cord will save a lot of headaches.

Watch your polarity! Attaching the power cord backward will do severe damage. Power cords have one red and one black wire. Red is positive.

---

[14] Radio Frequency Interference (noise).
[15] Twelve feet of wire is six feet from power supply to transceiver and six feet back.

# TRANSMITTER

The transceiver combines a transmitter and receiver in one box. When you select a transceiver, the transmitter section is not nearly as important as the receiver, and the receiver will dictate your choice.

Fortunately, this is not much of a dilemma. Most transceivers deliver 100 watts, and 100 watts are the same no matter where they originate. Perhaps one can argue about phase noise, clean signals, and key clicks, but modern equipment has (mostly) slain those dragons.

The vast majority of hams operate with 100 watts and, most times, that is sufficient power. It may not guarantee a place at the head of the line in a contest or DX pileup, but will provide communications over reasonable paths.

Some transceivers output 200 watts. Do you need 200 watts? Doubling power only increases the signal by 3dB, or one-half an S-unit. You can't hear a half S-unit difference, and your money should go toward a better antenna. Later chapters discuss antennas and amplifiers and considerations associated with them.

Bells and whistles drive up the price of modern equipment. If you are on a budget, just accept whatever the transmitter portion offers. The basic radios will suffice. Concentrate on buying the best receiver.

Helpful but non-critical, features include:
- Built-in antenna tuner.
- Built-in power supply (rare with modern equipment).
- Additional band coverage (6 meters and VHF/UHF).

- Built-in speech processor so you can tailor your audio.
- Built-in voice and CW keyer to play back pre-recorded messages, although you can do this with outboard equipment.

# RECEIVER

The receiver is the most critical part of any transceiver and where you should concentrate your attention when comparing radios.

Sensitivity is not the issue as they are all about the same and will receive at the natural background noise floor. What sets an excellent receiver apart from a mediocre one is selectivity, the ability to discriminate one signal. Selectivity to resist intermodulation by-products (IMD) from nearby strong signals is very important. IMD is an unwanted mixing of signals that sounds like growling or distortion.

One tool to defeat such interference is called a "roofing filter." The roofing filter gets applied very early in the receiver chain to reduce the breadth of signals reaching the later, more sensitive circuits. A narrow roofing filter will also minimize AGC[16] pumping caused by loud close-by signals.

Roofing filters or not, you need filtering down the chain. Crystal filters are one answer. Look for a receiver that accepts filters appropriate for the mode. Some low-end transceivers ship with one moderately wide SSB filter, around 2.7 kHz, and don't have a way to insert additional filters. Even if you don't order the narrower filters right away, it is nice to know you can add them later.

---

[16] Automatic Gain Control operates to level out the audio. AGC will automatically reduce the gain on a loud station to protect your ears.

## CHOOSING EQUIPMENT

Consider a narrow 1.8 kHz filter for SSB and a 500 Hz CW filter. New, these cost about $150 each, so a used radio that is "fully filtered" may be a bargain.

These are crystal filters. Another type of filter is Digital Signal Processing. Digital Signal Processing converts the signal to numbers (1's and 0's). Then, it looks for patterns, manipulating them to cancel interfering signals and noise. The math can be staggering, but microprocessors in the radio make it possible. These are the software-defined radios (SDR) mentioned earlier.

A crystal filter has a fixed width. Digital Signal Processing is continuously adjustable, and not limited to preset crystal-filter bandwidths. DSP processing was expensive, but is now found in modern entry-level transceivers. DSP effectiveness has advanced dramatically. Some manufacturers argue most users don't need roofing and crystal filters.

Another useful transceiver feature is the ability to operate in split mode, where a DX station transmits on one frequency and listens to another. This avoids people piling on the DX frequency, making him unreadable. It would be nice to hear both frequencies, so you hear the DX and the callers. Many transceivers have a VFO A and a VFO B but implement them differently.

An ideal set-up is a transceiver with two independent receivers so you can put the DX in one ear of your headphones and listen to the pileup in the other ear. You tune around on the second VFO listening in that ear for the station who just worked the DX, revealing where the DX is listening. With a transceiver set for split operation, your call will go out on the second VFO frequency, not the DX frequency.

An alternative to two receivers is sometimes called "dual-watch." Dual watch allows you to hear two frequencies but they are mono, so there is no separation of the DX and callers. Another alternative is to press a button and listen to the second frequency. The button-pressing rhythm is critical, or you'll miss the call. Two receivers are the best implementation, but that is more expensive.

Don't miss out on the fun of chasing DX. I hope you catch the bug and buy my DX book. EasyWayhamBooks.com

Norfolk Island, in the South Pacific, also has a connection to the Bounty mutiny.

# QRP TRANSCEIVERS

There is no set definition for QRP. Five watts or less is traditional but many consider twenty watts to be QRP. My first radio was a one-tube low-power transmitter and a military surplus single-band receiver. Primitive for sure, but it was a thrill to make a contact.

The minimal station is low power and, at the bottom end, CW only. There are many inexpensive low-power CW radios available. They are crystal controlled, so you operate on one frequency. Recently the Chinese have offered several versions for less than $15 on eBay and Banggood.com. They are kits and carry names such as Frog Sounds and the Forty-9er. I am not recommending such a minimalist approach as The Easy Way to make contacts, but it is certainly the inexpensive way.

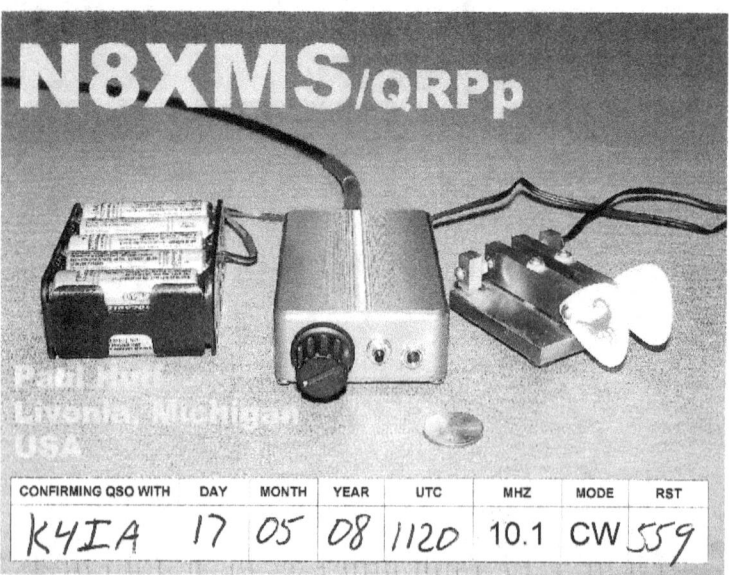

QRPp means very low power. This radio is crystal-controlled, CW only and runs a couple of watts.

More sophisticated low-power CW-only transceivers cover multiple bands and use a VFO[17] instead of crystal control.  Such radios cost $100 and up for a single band and $300 and up for multi-band.  Check out the later chapter on learning Morse code.

There are minimalist radios for digital modes such as PSK31 and FT8.  These digital modes are very robust and can provide world-wide communication with just a few watts and a simple antenna.

There are several low-power transceivers capable of CW, SSB, and digital operation in the under $400 price range with sophisticated models over $1,000.  Operating SSB at low power is a challenge.   Add an outboard amplifier to bring them up to 50-100 watts later if you are on a budget and want to build your station over time.

The point of this discussion is to emphasize that radio can be an adventure with any station.  Get on the air and start enjoying it.

---

[17] Variable Frequency Oscillator so you can operate on more than one frequency.

# PANADAPTERS

A panadapter shows a range of signals on a visual display. Panadapters feed the receiver's Intermediate Frequency (IF) into a Software Defined Radio (SDR) receiver and use software to show the results on a computer monitor. A computer logging program or a stand-alone solution provides the software to create the display.

The latest generation of modern transceivers feature a small screen on the face. Here is a panadapter displayed on a 23-inch monitor.

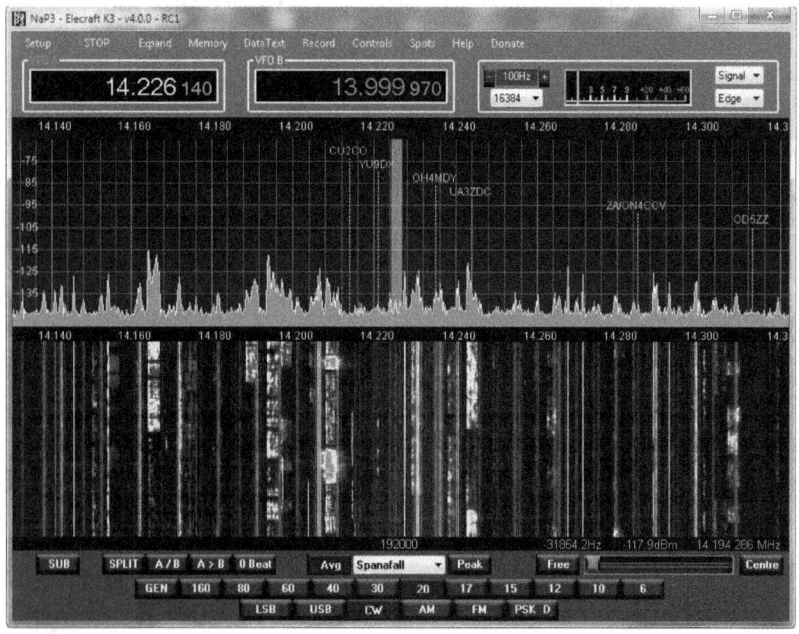

The middle of the screen is the spectrum showing signals across the band. The lower section is a waterfall. The trace moves down the screen with the newest signal at the top. Click on a signal and the radio QSYs[18] to it.

---

[18] QSY means "Change frequency."

One glance at the panadapter shows if a band has any activity and where the stations are. You don't have to spin a dial across the entire spectrum listening to find out. Panadapters are not a necessity, but are a great feature on modern transceivers.

# OTHER ACCESSORIES

The modern ham shack includes much more than a transceiver and antenna. There are dozens of accessories. You do not need all these gadgets, but here is a description of the more useful accessories.

**Antenna Analyzer** Measures SWR, reactance, and resistance of the antenna system. An analyzer differs from an SWR meter because the analyzer generates its own low-power signal and can display a range of measurements over different frequencies. The SWR meter uses the transmitter's power, only reads one frequency at a time, and only measures forward and reflected power, deriving from that the SWR.

**Antenna Switch** Handy for changing antennas and a lot easier than screwing and unscrewing coax connectors. Many provide a center "ground" position, but you shouldn't rely on that for lightning protection. See the later section on "Grounds and Lightning Protection."

**Anderson Power Poles** These are a standard plug-in power connector for attaching a power supply and equipment.

**Backup power** If you want to be ready for an emergency, you need a source of backup power. Generators are an obvious answer, but they can be loud and require a source of fuel. Solar power needs battery back-up.

## CHOOSING EQUIPMENT

Don't overlook your car as an energy source. The average car consumes one-third to one-half gallon per hour idling. A full tank of gas, run a few hours on and off to charge batteries, could keep you on the air for weeks.

Typical automobile batteries provide a great burst of power to turn the car's starter motor for a short time. Deep-cycle batteries, such as used with trolling motors, provide power over a longer period. Both are lead-acid batteries that discharge quickly and suffer permanent damage if discharged to 50% of capacity. Lithium Iron Phosphate (LiFePO4) batteries hold voltage longer (90%), weigh less and survive more charge cycles.

**Baluns and line isolators** If you experience RFI (Radio Frequency Interference), these may be the cure. See the Chapter on "Grounds and Lightning Protection."

**Clock for UTC** Coordinated Universal Time is the standard used throughout the world. We used to call it Greenwich Mean Time. Local time is too confusing when you communicate across time zones.

**CW Keyer and Paddles** "Paddles" refers to a two-part Morse code key. Push in one direction, the electronic CW Keyer generates a series of dits, and the other direction produces a series of dahs. Paddles can send faster than a straight key with less motion and fatigue. Your transceiver may have the keyer built-in, but you would need a set of paddles to operate it. Pietro Begali's QSL card in the Modes chapter shows a set of paddles.

**DC Power Distribution Panel** You might have several pieces of equipment to hook into your DC power supply. A distribution panel with multiple connectors makes it easier to connect them all.

Anderson Power Poles are the standard for convenience.

**Dummy Load**  A large resistor, sometimes air-cooled and sometimes in a bucket of oil.  You test or tune up into the resistor, so your signal does not go out over the air.

**Filters**  There are many kinds of filters serving different purposes.  AC line filters reduce voltage spikes and block noise on the AC line.  Audio filters can use tuned circuits or digital signal processing to reduce noise or narrow the audio passband to reduce adjacent interference.

RF filters suppress RF energy outside their design frequency.  If you have a powerful AM radio station nearby, an RF filter might help avoid it overloading your receiver.

**Headphones**   You don't need to spend a fortune on high-fidelity studio equipment because deep bass and tinkling highs don't matter.  The most important specification for headphones is that they must be comfortable and not crush your head in a vise.

There are three basic headphone designs for different occasions.  Over-the-ear models cover the ear entirely and block out external noise.  They can become uncomfortable if worn too long.

On-the-ear designs sit on top of the ear and don't provide much in the way of noise-blocking but are more comfortable.  I prefer on-the-ear unless I am in a noisy environment, such as a contest station where other operators are talking at the same time.

Earbuds fit in the ear and do a better job of noise blocking than on-the-ear designs and are more

## CHOOSING EQUIPMENT

comfortable than over-the-ear. I like earbuds for long sessions.

**Magazines** The two preeminent ham Radio print magazines are *QST* and *CQ*. *QST* is the official journal of the ARRL.[19] *CQ* is less formal and less technical but more fun. I enjoy them both. Both are available in a digital version. There are several on-line eZines. The most prominent are eham.net and QRZ.com.

**Multi-Meter** These handy devices measure volts, amps, and resistance. Every ham should have at least one. The cheap versions (under $8) are not laboratory-grade but are sufficient for most of our uses. Harbor Freight often offers them free to get you in the store. I make it a point to pick one up whenever I can.

**Speaker** The speaker in your transceiver is probably not ideal. It fires upwards instead of at you. It is also small. Wide fidelity is not essential and the built-in speaker might be sufficient. An external speaker could improve your reception. You don't need to pay a premium price to get a speaker that matches your rig. When they say the speaker matches, they only mean it is painted the same color. See the chapter on Shack Design.

**Speech Processor** A speech processor tailors the audio to increase readability. It adjusts the frequency response to emphasize mid and mid-high frequencies that carry the most information.

**SWR Meter** SWR, or Standing Wave Radio, is a measure of the match between the antenna system and the radio. The SWR meter goes between your transmitter and the antenna to measure forward and reflected power while transmitting. Less reflected

---

[19] American Radio Relay League bills itself as the "voice of amateur radio."

power indicates a better match. Analog ones have two needles. One needle measures forward power, and the other reflected power. A scale where the needles cross, shows you the SWR. Digital meters, often found in automatic antenna tuners, might display all three measurements on a small LED screen.

**Voice Keyer** A voice keyer allows you to record a message and play it back with the push of a button. This is very handy for calling CQ. Set the message to auto-repeat, mash the button, sit back, and wait for an answer.

The best accessory is an assistant, as shown on this card from the Canary Islands.

# TRANSCEIVER FRONT PANEL CONTROLS

There are an overwhelming array of controls on a modern transceiver. Large panels allow room for lots of knobs and buttons for your adjustments. Small transceivers have very few controls and rely on multiple button pushes or menus to dig down into the parameters you might adjust. Here is a list of standard controls and what they do. I use the terms "switch" and "button." One you flip, and one you push, but the function is the same.

**AF Gain** A volume control for the audio amplifier. Compare to the RF Gain control for the radio-frequency stage amplification.

**Antenna Selector** A transceiver may have more than one antenna jack on the back. Select the antenna from this switch or button. If the receiver seems dead, check that you chose the right antenna.

**AGC (Automatic Gain Control)** Suppose you are listening for a weak signal and have the gain turned up. Then you tune across a very loud station. AGC levels the volume to protect your ears.

AGC usually has two settings, fast and slow. "Fast" will recover quickly while "slow" will hold down the amplification longer. Use different settings depending on the conditions. Voice communications usually sound better with "slow."

**Attenuation** Attenuation reduces all energy coming into the receiver, both noise and signals. It is handy on the lower bands, where noise can overwhelm the sensitive circuits. Reducing the noise will make a signal easier to hear.

**Antenna Tuner** The antenna tuner, properly called a conjugate matching device, adjusts the load of the antenna system to match the transceiver. . Relays clatter as an automatic tuner works to find the right solution. Don't worry; it is not broken.

**Band Switch** Rather than spin the frequency knob to go from 40 meters to 20 meters, one push changes the band, usually returning to the last frequency you used on the new band.

**Compression** Your voice has loud and soft inflections. Some words are naturally louder than others. Compression works to fill out the audio by boosting the less-strong words and syllables. Adjust the amount of compression to get that boost, but not so much as to add distortion.

**CW Speed** Most transceivers have a built-in keyer. This control adjusts the speed of the Morse code sent from paddles.

**CW Pitch** The pitch is the musical tone. If you have it set to 650 Hz, when you hear the received audio signal at 650 Hz, you will be on the same radio frequency as the sender or "zero-beat." Pick a tone you like to hear and teach your ear to remember it.

**Frequency Entry** Enter the frequency on a keypad rather than spin a dial. It may require an additional button push (Enter) to activate the change.

**Filter Width** This may be fixed or variable. Narrower filters restrict the range of signals heard.

**Filter Shift** (PBT, passband tuning) Alters the shape of the filter to favor frequencies above or below the center of the filter.

## FRONT PANEL CONTROLS

**Mic Gain** Adjusts the volume of the microphone input to the transmitter. Talk close to the microphone and adjust the mic gain per your transceiver's instructions. Most hams "close talk" the microphone within an inch of their mouth. If you speak too far away and turn up the mic gain to compensate, it will also pick up background noises such as barking dogs, TVs, or fans.

**Message Playback** Some transceivers have a built-in recorder (voice-keyer) for voice and CW play-back. Record a message and play it back with one button push. A recorded message makes it a lot easier to call CQ and helps cure mic fright. Pushing a button is less intimidating than calling.

**Memories** Memories store frequencies and modes. They make it easier to return to a particular frequency and are handy to check into a net. Set a memory for the top, middle, and bottom of a band to move around without spinning the dial.

**Mode** Selects AM, USB, LSB, CW, and Digital modes.

**Monitor** Listen to the transmitted signal through this function. Keep the monitor volume low, so your ears stay sensitive to receive. It is probably not needed on voice modes.

**Noise Blanker** Activates a circuit designed to mute the sound of repetitive noise such as a spark plug or electric fence.

**Noise Reduction** Filters out random noise. There may be multiple settings to deal with differing conditions. Very aggressive noise reduction can introduce distortion. This adjustment is by trial and error and changes with conditions.

**Notch**  Reduces or "notches" out an offending signal. Auto-notch will seek and silence a carrier[20] near your frequency.  Adjust a manual notch to do the same thing.

**Power**  The on/off button, of course, but there is also a control to set how much power to transmit.

**Pre-Amp**  Turn on an additional amplifier applied to signals before they enter the receiver.  Use this sparingly and only on higher frequencies with low background noise, or it will overwhelm the receiver.

**QSK**  When activated, the transceiver will switch from transmitting to receive almost instantaneously.  QSK is very helpful for CW to hear the receiver between sending characters.

### Receiver Incremental Tuning (RIT or Clarifier)
Tunes the receiver without changing your transmit frequency.  If someone calls slightly off frequency, use this to zero them in and "clarify" the signal.

If you adjusted the VFO to make the incoming signal sound better, it would change the transmit frequency and sound "off" on his end.  Then he would adjust, and the two of you would chase each other all over the band.  Transmit in one place and use the RIT to correct for the off-frequency signal.

**Reverse**  To change from USB to LSB or to change the side on which you hear a CW signal.  This is an effective interference fighting tool on CW.  Reversing the side will make the interfering signal further away from the tuned frequency.  On SSB, stick with the convention for the band.  It does no good to listen to

---

[20] If you are listening to SSB and someone decides to tune up near your frequency, you will hear the tone of his carrier.  Notch will cut it out.

## FRONT PANEL CONTROLS

an upper sideband transmission on a lower sideband setting. You won't be able to understand it.

Another use of term "reverse" is to exchange VFO A and VFO B frequencies. That is a different control. (See below).

**RF Gain** Adjusts the gain in the radio-frequency section of the receiver. Turn the RF Gain down to barely hear the background noise. This setting preserves the maximum dynamic range of the receiver. If a signal is below the noise, turning up the noise won't make it any easier to copy. Turning down the noise is also less fatiguing.

**Spot** Injects the audio tone selected with the CW pitch control to help zero-beat the other station.

**Squelch** Silences the receiver when there are no signals, the same as on an FM HT. Don't use squelch on HF because it only triggers with signals substantially above the noise level. HF signals close to the noise level won't be loud enough to open the squelch.

**Transmit Incremental Tuning** Changes the transmit frequency while leaving the receive frequency the same.

**Tuning step** Determines how quickly the frequency changes as when the dial spins.

**VFO** Variable Frequency Oscillator is the frequency tuning control.

**VFO A / VFO B** If the transceiver allows, you can have two frequencies active. Transmit on one, but in split operation, listen to the other. The **A/B** button would switch the two.

**VOX**  Engages the voice-operated relay, instead of using a transmit/receive switch like the push-to-talk. VOX facilitates hands-free operating.

**VOX Delay**  Sets how long the radio stays in transmit after you stop talking.  It can be very distracting if the delay is set too short.

**VOX Anti-Vox**  Noise from the receiver speaker can fool the VOX into thinking you are talking and want to transmit.  The AntiVox control is negative feedback that cancels out the received signal, so it doesn't trip the VOX into transmit mode.

GRID: EL 88  **HILLSBOROUGH COUNTY**

Frank Moore
11801 North 51st Street
Temple Terrace, Fl 33617-1403

This QSL card shows the front panel of an Elecraft K2 transceiver.  It features multi-use buttons.  Push activates one function.  Push-and-hold activates a second.

# TRANSCEIVER REAR PANEL

The rear of a transceiver can be as confusing as the front. In addition to input and output jacks, there may be controls for functions that don't often change, such as microphone compression or VOX delay. If you're looking for a control and can't find it, check the back of the radio. Also, look for a trap door on top of the chassis.

The rear contains the connections, the goes-intas, and goes-outtas. There could be discrete jacks and cables for various functions, or they might be bundled in a DIN connector. "DIN" stands for Deutsches Institut für Normung, the German national standards organization. A DIN connector is small and round with 5 or more pins. The plug has a key tab, to only fit the jack in one orientation. If there is a problem pushing it in, you probably don't have the key tab lined up correctly. Don't force it!

The rear of the equipment might be hard to see, or the only way to see it is to lean over the front, and then everything is up-side-down. Take a picture or sketch to make changes easier. I put up-side-down labels on the rear so I could read them hanging over the top. Those connectors all look alike otherwise.

Here's a list of what you might find on the back of a transceiver:

**Accessory Jack** Control other equipment or pass audio. The various pins may allow for additional audio inputs and outputs for packet or RTTY signals.

**ALC** Automatic Level Control interfaces with an amplifier to assure the transceiver doesn't apply too much power and overdrive the amp. If the amplifier senses it is being over-driven, it changes the ALC voltage to reduce power from the transmitter.

**Amplifier Keyer** (may be called "key out"). A cable connected to an external amplifier to switch it to transmit mode when the transceiver is sending.

**Antennas**. Jacks for the HF and VHF/UHF antennas.

**Fuse or circuit breaker** In addition to a fuse in your power cord, many transceivers have a safety fuse or breaker. Keep a few spare fuses in your toolkit.

**Grounding lug** To attach the station grounds. See the chapter on "Grounds and Lightning Protection."

**Key or paddles** Input for CW operation. A key uses a mono plug for its two wires. Paddles use a stereo plug for ground, dits, and dahs.

**Line in and Line Out** Fixed-level audio for digital modes and voice keyers or recorders. These may be a DIN plug or individual audio plugs.

**PTT** Separate push-to-talk switch. The microphone may have a button that activates through the microphone jack. A separate PTT connector is for a footswitch or other hand switch.

**Serial/USB port** for the radio and computer to communicate. Older Icom radios use a CIV interface that looks like a 1/8 inch phone jack.

**Speaker output** Plug in an outboard speaker.

# AMPLIFIERS

Some folks work QRP (low power). My hat is off to them. I have done it myself, including this two-way QRPp (really low power) QSO with LZ2RS in Bulgaria. Rumi and I were both running 1/10 of a watt which calculates to 49,500 miles per watt. That may be a fun way, but it is certainly not The Easy Way.

Most hams getting started on HF are not looking at amplifiers and accept 100 watts as sufficient. Make no mistake. 100 watts or less will work the world.

Want more power. How much is enough? Remember that doubling power adds 3 dB to the signal or one-half S unit. Going from 100 to 200 to 400 watts adds one S Unit. 400 to 800 is another half, and 800 to 1600 only adds another half. Can you hear the difference between S6½ and S7? No, you cannot.

There are many tube and solid-state amplifiers in the 500-800 watt range that will serve well and not break the bank or your station. High power requires upgraded switches, coax, and antennas. Mid-range amplifiers should give plenty of power. 500 watts is more power than most. Some countries limit their operators to 400 watts.

Depending on the rest of the circuit load, 500 watt amps may run from a 110-volt power outlet. That saves the extra work needed to get 220 volts to the shack. I was fortunate that the electric dryer outlet was accessible nearby, and I could tap into it to get 220 volts. Run the higher voltage if possible.[21] Your amplifier will appreciate it. Ohm's Law says the current draw at 220 volts will be half that at 110 volts. The wire's power loss is one-quarter, making it less likely to suffer component-stressing and distortion-inducing voltage sag.

The choice of tube vs. solid-state usually boils down to price. Solid state can be twice the cost per watt. The pros for solid-state amplifiers are instant on, no twiddling of knobs to tune, and instant band changes. The cons are they do not tolerate SWRs over 2.5:1, so require an outboard antenna tuner if the antennas aren't perfect. Manual tuners require knob twiddling and high-power auto tuners can be expensive.

Tube amplifiers are much more tolerant of misadjustment. The large metal surfaces in the tube are not so fragile as the small junctions in a transistor. Solid-state is considerably more expensive than tube, and a blown transistor is harder to replace than a blown tube.

I've had both, and there is no "right" answer for this one. Maximize the return with a moderate tube amplifier and put the money saved into a better antenna. A better antenna will make a difference on both transmit and receive. Don't be an alligator – "all mouth and no ears." More on antennas later.

---

[21] I tapped into the nearby 220-volt dryer circuit. It might not have been "to code," but it saved running a line all the way to the breaker panel.

# SHACK DESIGN

Make the shack comfortable and inviting.  I shudder to see hams relegated to windowless, damp and dingy basements crouched among the cobwebs, behind the furnace, out of contact with the rest of the household. Radio is your pride and joy, so come out of the cellar. Spouses don't hate your hobby. They hate your isolation.  Be part of the family.

Worried about making a lot of noise in the living area? Digital modes can operate silently, and headphones confine CW.  Find a spot somewhere on the main level to get away from the TV and have a little solitude.  Is there a den or sun porch?  I like to look out the window and see the bird feeder or the grandkids.

No matter where, brighten it up with adequate lighting and put a rug on the floor.  Get a comfortable desk with enough room to spread out.  I used a door on sawhorses as a kid.  Now, office furniture is available in comparatively cheap, attractive, and utile packages. Get a decent office chair and stop sitting on that old unpadded and broken kitchen stool.  Make yourself comfortable and be a part of the family.  Maximize radio time without killing your attitude, your posture, or your marriage.

Since the computer has become an essential element in the modern ham shack, the monitor is front and center.   Multiple 23-25 inch monitors will give more square inches of screen than one large monitor.  I have four.  I originally had them stacked but found it was very uncomfortable to look up at the higher monitor, so I changed them to side-by-side.

Things that need adjusting should be within easy reach.  I am right-handed, so my transceiver is where I can twiddle the knobs with my right hand.  The radio belongs on the desk level to rest my arm while tuning.

Having the radio on a shelf above the desk is very tiring and uncomfortable. My key is also to the right, along with my solid-state amplifier.

I have two small Pyle® speakers on either side of the radio. They provide excellent stereo. Many transceivers allow for phase reversal or other audio effects that enhance the sound, giving the impression the signal is in your head, not your ears. When the going gets tough, I pull out the cans (headphones).

Computer monitors dominate in my shack.

Bookshelves in the background hold auxiliary equipment such as the antenna tuner and wattmeter; all of which are eye level and within reasonable but not immediate reach. I don't need to adjust those components often, but need to see the displays. The bookshelves were too tall, so I cut about six inches off the bottom and rearranged the shelves.

As a general rule, anything that does not need adjusting can be out of reach. The antenna rotor-control box is in sight but out of reach. A rotor control application on my computer turns the rotor, so I don't need to spin the dial manually.

The power supply is on the floor, where it belongs. I use my big toe to turn it off and on, but most of the time, I leave it on. A power supply next to a transceiver can induce hum and other noise. It also takes up space at the operating position. You never adjust the power supply, so it doesn't need to be on the desk.

An L-shaped desk makes it easier to place two monitors, as seen in this Irish station.

# COMPUTERS

Radios and computers go together, like peas and carrots. The modern transceiver interfaces with computer logging programs, digital modes, CW keyers, voice keyers, rotor controls, panadapters, Internet spotting, call sign lookups, and QSLing. The computer has become the central control point for a modern ham station.

There is some software written for Macs, but the majority is for PCs. Very little runs on Linux. Chromebooks rely on access to the Internet to run programs, and I am not aware of any ham software that will run on a Chromebook. There are some APPS that will work on Apple and Android devices to run PSK and simple logging. Experimenters are always thinking expansively.

The computer does not have to be expensive and high-powered. Add some memory by visiting Crucial.com and running their memory scanner. Adding memory is the most cost-effective way to improve computer performance. Ham programs don't need a ginormous hard drive or an exceptionally fast processor.

I prefer a desktop over a laptop because it allows me to upgrade components. There is no reason for it to be on the operating table. Bond the case to the shack ground and other equipment for lightning and RF protection. Check out the next chapter on Ground and Lightning Protection.

An Internet connection downloads spots, looks up callsigns, backups your log and uploads logs to LOTW[22]. A slow Internet connection is sufficient as these are not intensive uses.

---

[22] Logbook of the World. See the chapter on Confirming Your Contacts.

## COMPUTERS

Newer radios communicate with the computer's USB port, but older radios use a serial port interface. New computers don't come with serial ports, but cheap serial port cards plug into the motherboard. There are also USB to serial converters to feed the radio. ProLific and FTDI converters have had issues with counterfeit chips. Both companies inserted code in their drivers to disable unlicensed fakes. Do not buy cheap knock-off converters and cables.

A computer can send digital modes such as PSK or FT8 and also act as a voice or CW keyer playing pre-recorded messages. On voice or digital modes, sound from the computer feeds into the transceiver as described in the chapter, Modes. For CW, an interface turns computer-generated pulses or codes into Morse code.

There are many devices designed to provide isolation and trigger your PTT. Check the "Interfaces" reviews on eham.net.

# GROUNDS AND LIGHTNING PROTECTION

There are three "grounds." An electrical safety ground, a lightning ground, and an RF ground. They fulfill three different functions but must operate in harmony and tie together. The fundamental concept is to avoid any difference in potential between components, and that means minimal resistance to both AC and DC.

## ELECTRICAL SAFETY GROUND

The electrical safety ground is the third hole in an electrical outlet. The plug wire connects to the case of the equipment, whether that equipment is a radio or a dishwasher. The outlet's third hole connects to the green wire flowing back to the electrical panel.

We call this a "safety" ground because the purpose is to trip the circuit breaker if voltage appears on the case of your equipment. The safety ground protects against electrocution.

Do not defeat the safety ground protection with an adapter plug or any other gimmick!

## LIGHTNING GROUND

The lightning ground deflects power from a lightning strike by dissipating the energy before it enters the house. Thor's hammer carries an incredible wallop. A thunder cloud can hold over 100 million volts of potential power and typically generates 5,000 to 20,000 amperes of current. A direct hit will be devastating.

Lightning also creates an electromagnetic pulse. An indirect or nearby hit can induce damaging amounts of current into your antenna and equipment. I had the

## GROUNDS AND LIGHTNING PROTECTION

front end of a transceiver destroyed by the pulse from a nearby strike. Five surface-mounted parts, each the size of a flea, needed replacing.

Reduce the potential between, around, and through your equipment and your home's electrical system. Search for and download the free Polyphaser "Lightning Protection and Grounding Solutions" PDF book. The book includes robust solutions commercial installations use to survive direct strikes.

Give lightning a path to the ground outside the home. A lightning arrestor in the feed line and attached to a ground rod is just a start. Multiple 8-foot ground rods, connected in parallel, are better than one.

Lightning arrestors provide some protection against nearby strikes and may bleed off potential charges to prevent a strike. The arrestor has a gap that will allow some of the potential on the conductor (coax, ladder line, or rotor cable) to jump to ground. A few nearby strikes can damage the gap. Other designs use a coil to provide a DC ground. The coil has a high impedance to RF. Inspect and replace lightning arrestors often.

Note the words "some protection" used above. Each connection and each piece of equipment has resistance and is, therefore, at a different potential. Voltage applied through differing resistances causes current to flow. Even if one component has a very high resistance, some portion of the current will flow through it. High voltage can induce a lot of amperes.

Putting an antenna switch in the "ground" position is not enough. The antenna may be "grounded," but all the other cables coming into that switch are connected, and current will flow back, to, and through the connected equipment.

Don't rely on a lightning arrestor or switch for full protection. A direct hit will destroy the arrestor. Disconnect an unused antenna completely. Unscrew the coax[23] and don't leave it lying on any equipment or near a path to ground. That lightning bolt travels thousands of feet and can easily jump a few more to get to your equipment or electrical outlet.

Summer thunderstorms are not the only source of damaging energy. Static electricity generated by wind or rain can also cause damage. Dry winter air encourages static charges on the antenna. The components in receivers are designed to handle millionths of a volt not millions of volts. Cultivate the habit of disconnecting antennas when not in use during all times of the year.

As an additional precaution, unplug the power supply and other equipment to protect against a power surge on the AC line. I lost a very expensive amplifier to a power surge. Even though the amplifier was "off," the power supply remained energized. When the lights flickered, the power supply fried. (See the Chapter on Insurance).

# RF GROUND

We often think of "ground" in terms of direct current or alternating current at house frequencies, 60 Hz. A cold-water pipe may provide a safety ground, and may be of limited help to dissipate lightning surges, but it is not an effective RF ground.

"Ground" is not some mysterious pit in which you dump energy. At radio frequencies, every piece of wire has inductance and capacitance. The wire presents some impedance, which is AC resistance.

---

[23] You can also find push-on coax connectors for about $5. They are easier to connect and disconnect.

## GROUNDS AND LIGHTNING PROTECTION

A ground wire that is one-quarter wave long presents an infinite impedance to the equipment leaving the equipment ungrounded, and worse, at a high potential above ground. That is because a one-quarter wavelength line acts as an impedance transformer, and a short to ground at one end will be open at the other. A one-half wavelength ground line will mirror the impedance at the end, and a short to ground would appear as a short to ground at the equipment as well. Every other length will present some less-than-ideal combination.

A good RF ground is a low impedance path for RF energy to flow around and not through equipment. RF energy also causes "mic bite," a burning sensation when you touch the microphone or chassis while transmitting. It might also cause distortion on your SSB signal. A computer might reboot for no reason, or the radio display could go haywire. Thee might be excess RF floating on the equipment even with these extreme symptoms.

There are lots of ways RF can get in the shack. One is an unbalanced antenna system. Even a balanced antenna such as a dipole can become unbalanced because of objects within its field, such as uneven terrain or metal siding. Another cause might be a feed line that is not perpendicular to the antenna as when the feed line is closer to the antenna on one side. Balanced feed lines cancel RF, and unbalanced feed lines radiate.

Another troubling source is RF induced by the antenna into wires within your shack or house. The house mains, telephones, alarms, televisions, and TV cable all have wires which act as antennas. Shack components such as the computer keyboard, mouse, video display, microphone, coax, and speakers use wire. Wires receive like an antenna, bringing RF wherever they connect. Currents on the wires will also radiate, causing RFI in nearby equipment.

There are three ways to control RF on wires and feed lines. Proper planning uses all three. First is a choke, a coil of wire that impedes RF (inductive reactance). Increase the effect by wrapping the coil around a ferrite mix of ceramic and metal. There are different metal mixes chosen for the frequency you are trying to choke. Mix 31 is used below 10 MHz. Above 10 MHz, mix 43 or 61 prevails.

A toroidal choke is donut-shaped. Wrap the wire through the choke as many times as will fit. The impedance goes up with the square of the number of turns. Going through one choke two times yields four times the impedance. Go through two chokes in series only provides twice the impedance. More turns are better than more chokes.

To prevent RF on your antenna's coaxial feed line, use 10-13 turns of coax around a six-inch-diameter-PVC pipe. Products marketed as line isolators and isolation baluns are also chokes.

Here is an excellent article by K9YC, Jim, dealing with RFI and chokes. It should be required reading: audiosystemsgroup.com/RFI-ham.pdf

The second solution to RF induced in wires is twisted-pair wire. The twist will self-cancel any induced currents. Choose twisted-pair for equipment interconnections.

The third solution is bonding, connecting equipment cases together. Bond each case to the others to provide a low impedance path for RF around the equipment rather than through the equipment and interconnected cables. If RF has two paths, one through your equipment and one around the equipment, the current will split among the parallel paths.  Most will follow the lowest-impedance route.  Provide an alternate low impedance (resistance) path that keeps the RF out of the circuits inside your radio.

Connect all chassis: radio, antenna tuner, computer, and amplifier with a ground strap to provide a low impedance path from chassis to chassis.  That encourages RF to stay off the interconnecting cables and out of the electronics.   Also, run a strap to your shack's common ground connection.  Use bonding strap parallel to any cables or coax running between components.  The result is multiple connections among the equipment and station ground.

RF travels on the surface of a conductor because of a phenomenon known as "skin effect."  Ground strap is flat, not round like wire, so it has more surface area than round wire and is a better RF conductor.  Skin effect means the strap can be thin.

Strap comes braided or solid.  Copper roof flashing works but has sharp edges, so be careful while handling.  Solid flat strap is a better conductor than braided. Braided is more flexible but more likely to corrode when used outdoors.

To attach the strap to a chassis, poke or drill a hole through the material.  Sandwich the strap between a couple of large washers and secure with the nut on your equipment's ground lug.  If you need to splice straps, use the same technique of sandwiching them between two large washers.

Finally, electrical codes require you to bond the RF, safety, and lightning grounds together.  There should be no difference in potential between the various parts of the grounding system.  Any difference in potential will cause current to flow between the components bringing hum, buzz, RFI, or damaging current from lightning.

ARRL's book "Grounding and Bonding for the Radio Amateur" is also excellent.

The only thing more dangerous than bad grounding is crocodile taunting, as seen here in Chad, North Africa.

# ANTENNAS

There is probably no more contentious area of debate than the choice of antennas. "If you ask five hams a question, you will get seven different answers and maybe a fistfight."

A better antenna helps both transmit and receive, so money spent on the antenna system is very beneficial. Improving an antenna system is the single most important investment. For example, a two-element yagi with 100 watts is about the equivalent of a 400-watt amplifier to a dipole. The yagi is cheaper, has the same multiplier effect on receive, and attenuates signals coming from other directions.

I don't expect you to start with a yagi, and you may never get to erect a monster tower equipped with stacks of mono-banders. Do the best you can and get on the air. Who cares if you are an S unit or two below the strongest station on the band? The proper operating techniques will get plenty of contacts.

There are thousands of antenna designs and hundreds of books and articles on the topic. I am showing The Easy Way to get on the air without spending a lot of money and with minimal complication. The simple antennas discussed here will fulfill those requirements. You can (and will) spend the rest of your life building, designing, and experimenting.

Don't succumb to paralysis by analysis. **The antenna you have up will work a lot of stations. The perfect antenna you are looking for, but still haven't installed, won't work anyone.**

# THE HUMBLE WIRE DIPOLE

A single-band wire dipole is simple and cheap. Measure and cut two wires[24] according to the formula 234/frequency = length in feet. Appendix A is a chart for each half of a dipole.

That distance is the length of wire from the center insulator to the end insulator. You need to add about 18 inches, so you have 9 inches on each end to wrap the wire back through the center and end insulators and around itself. Removing excess wire is a lot easier than soldering on a longer piece. Cut long.

Attach one end of each wire to each side of the center insulator. Feed in the middle with coax by soldering the center conductor to one side, and the coax braid to the other. Affix the coax to the insulator with some form of strain relief. The soldered connections can fail if stressed by holding up the weight of the coax. Pre-made center insulators with a standard PL259 coax jack save the bother of soldering and provide strain relief.

Intruding moisture is coax's number one enemy. Seal the end of the coax and any connectors to keep out moisture. Use good quality electrical tape and cover it with Coax-Seal®, a putty-like tape. Then cover the Coax-Seal with another layer of electrical tape. The first layer of tape makes it easier to remove the Coax-Seal later.

Put another insulator on each end of the wire and attach the hanging rope to the end insulators. UV resistant antenna rope, sometimes called "Para-Cord," is best. Sunlight will destroy other materials in a few

---

[24] Use antenna wire. It is either copper-coated steel or has been stretched (hard drawn). Soft copper house wire will stretch considerably and change the frequency of the antenna. It is also not as strong.

months.  I have had antenna rope up for over ten years without a problem.

Then, hang as high as you can.  A center support, to relieve tension on the wire, is good but may not be necessary unless there is a heavy balun or lots of coax in the middle of a long wire.

Expect some sag in the middle.  Don't try to pull the wire taut, it may break.  Sag in the wire is called a catenary, and solving the mystery of sag vs. weight vs. wire strength is a classic calculus exercise beyond my pay grade.  There are calculators on the Internet.

When hanging an antenna in trees, leave some extra slack for the wind.  One trick is to put two loops in the hanging rope about six feet apart.  Tie them together with light line.  That line will break when stressed, and the hanging rope will unroll, relieving the stress.  Put it back together after the storm.

If there is not enough room for a flat top, stay horizontal as far as possible and let the ends hang down or zig-zag a bit.  Most of the radiation is from the high-current center of the antenna, so what happens out at the ends is not critical.

If there is only one support, spread the legs, the wider, the better.  Try to make the center angle at least 90 degrees.  This configuration is an "Inverted Vee."

I made it sound so simple when I said, "hang as high as you can."  Imagine standing on the ground looking up tree and wondering, "Just how am I going to climb up there?"

Relax, no climbing required.   The trick is to get a pilot line over the tree and use the line to pull up the hanging rope.  Use a drone or shoot a weight carrying the line over the tree with a slingshot or spud gun. Bow and arrow is a dangerous choice. I prefer about 20 lb test for the pilot line. Anything less might snap too easily. Heavier, and it won't break if you need to when the line gets tangled in a tree. Once the fishing line is over the tree, use it to pull up a stronger string such as mason's line. Then use the mason's line to pull up the heavier antenna and end rope.

There are commercial products available such as the EZ Hang slingshot or various pneumatic antenna launchers (spud guns or air cannons). A fishing reel handles the line. You can make your own, and there are plans on the Internet. I built a spud gun from the March 2009 *QST* magazine. Everyone in the club borrows it. Perhaps, someone in your local club has an antenna launcher to borrow for an afternoon.

Warning: Spud guns and slingshots might be considered "weapons" in your jurisdiction. Be careful and stealthy. Spud guns are loud.

If questioned, do not refer to the device as a spud "gun." The word "gun" will freak out the listener, especially a police officer. Refer to it is an "antenna hanging device" or "tree climber's aid." You are not "shooting" the line, you are "hanging" it or "placing" it. Rehearse these answers or be prepared to take the Fifth.

Trim the antenna for minimum SWR, but don't obsess if the SWR is below 1.5:1. Even 2:1 and higher is acceptable if your radio does not fold back power. If the SWR goes up with a higher frequency, it means the antenna is too long, wrap a bit of the wire back onto itself. If the SWR goes down with a higher frequency, it means the antenna is too short. Unwrap some wire.

## ANTENNAS

Appendix A is a chart for various frequencies.  If the lowest SWR is at 13.9 MHz (202 inches) and you want to move it up to 14.2 MHz (198 inches), the chart tells you to reduce 4 inches from each leg of the dipole. The easy way to do that is to pull 4 inches of wire through the insulator and wrap it around itself at the end of the antenna.  It is easy to shorten an antenna that way.  It is harder to add wire, so start a little long and wrap the excess back.

The antenna will move up in frequency when raised higher, so don't check the SWR on the ground and expect it to stay the same in the air.

Assuming a choice, which direction should the antenna hang?  Horizontal wire dipole antennas radiate off their sides, and that is where they have gain on the fundamental frequency.  If the antenna points north/south, it will radiate the best east/west.  A multi-band antenna on frequencies above the fundamental  develops lobes at angles to the sides. The lobes make the choice of direction a bit hit-or-miss, but for most of the US, I would say a north/south orientation is roughly best.

The pattern for horizontal antennas hung less than a half-wavelength high becomes omnidirectional.  The signal goes in all directions at a higher angle of radiation.  Inverted Vees are more omnidirectional than flat-tops.

Since you can't move the trees, the answer is to put the antenna wherever it fits.

# MULTI-BAND WIRES

Multiple single-band antennas require lots of wire and trees. There are many designs for multi-band wires to choose.

I have used a 136 foot off-center-fed (OCF) Windom that covers 80-10 meters. The OCF Windom is a dipole fed 20-30% from one end, off-center. The antenna has approximately 200-ohms impedance on the fundamental frequency and even multiples. A 4:1 balun transformer in the center transforms that to match the 50-ohm coax running to the shack. Thus, an 80-meter OCF will be a reasonable match on 40, 20, and 10 as well. You can force it to work on other bands with a wide-range antenna tuner.

One caveat for OCF antennas. Since the two sides are of different lengths, the antenna is unbalanced. Current can flow on the coax leading to RFI, radio frequency interference, or RF feedback. If you feel a burn when you touch the radio, that is RF on the chassis. Create a coax choke by wrapping 10 to 13 turns of coax on a 6-inch PVC pipe outside the shack.

Other multi-banders include fan and trapped dipoles. A fan dipole is multiple single banders fed from a common center. They can be tough to tune because the elements interact. Trim the lowest frequency first. Two dipoles for 80 and 40 meters work well on many bands.

Trapped dipoles include tuned circuits (traps) that break up the wire length at different frequencies. Fans and trapped dipoles should not require an antenna tuner.

## ANTENNAS

The classic multi-band wire antenna, called a doublet, is a dipole fed in the middle with ladder line[25] and tuned with an antenna tuner. How long? Generally, one-half wavelength on the lowest frequency but there are many variations. 102 feet or more will work on 80 meters and above. 88 feet will serve for 40 and above. That isn't a half-wave on any band. Those lengths are a compromise to get the best results on multiple bands and will require a wide-range antenna tuner.

What is an antenna tuner? Properly, we should call it a conjugate matching device because it matches the complex impedance of the antenna system to the transmitter. Matching assures that maximum power transfers to the antenna. It would be a rare case for an antenna to be a perfect match.

Many transceivers have an antenna tuner designed to correct SWR of 3:1 or less and may not have enough range to find a match for multi-band use. External antenna tuners usually handle greater mismatches and make the system work.

Just because the transmitter sees a matched load doesn't mean the system is without losses. SWR is the mismatch at the antenna. High SWR causes reflected power to be attenuated in the feed line. The worse the match and the higher the frequency, the more loss.

Many multi-band antennas use ladder line feed because the loss in ladder line is only 10% that in coax. Even though there is a mismatch from the antenna, much less power is dissipated in line.

---

[25] Ladder line is a transmission line made up of two parallel wires held apart by spacers. It has very low loss.

Ladder line has an impedance of 450 - 600 ohms. Many antenna tuners include a 4:1 balun to obtain a better match and to convert the balanced ladder line to unbalanced coax.

The downside of ladder line is that it needs to be kept away from metallic gutters or siding that would unbalance the fields around the wire. It also can't be buried. If you cannot run ladder line to your tuner, install a balun outside and run coax from the balun to the tuner. A short run of coax won't add significantly to losses.

The multi-band doublet is a handy, time-tested and simple antenna

Another multi-band variation is called the G5RV, after the callsign of the English inventor. The G5RV is 102 feet long and fed in the middle. The feed is a section of ladder line about 30 feet long and then converts to coax. That section of ladder line acts as a transformer and allows the antenna to achieve a reasonable match on most bands. The antenna tuner in your transceiver may be sufficient to make it work without a wide-range antenna tuner.

You can buy pre-made Windoms, G5RVs, doublets, and dipoles or roll your own. Plans are all over the Internet and in antenna books. I recommend stranded antenna wire, insulated or not. Copper plated steel is very stiff and hard to handle. Avoid house wire as it is pure copper and will stretch, moving the resonate frequency lower over time.

# VERTICALS

Not everyone has trees or other places to hang wire. Never use a utility power pole as it carries lethal voltages. The vertical antenna is an answer.

## ANTENNAS

Critics say verticals radiate equally poorly in all directions.  Yet, I worked tons of stations with a vertical on my roof.  My yard at the time was too small for anything more.  Verticals work best on low-angle radiation, like the signals that come in from far-off DX.  A vertical would not be a good choice to make contacts a few hundred miles away.

A properly erected vertical can be a good antenna.  It is relatively stealthy to answer concerns about neighbors and restrictive covenants.  Not only is it low profile, but you can also use a tilt-over ground mount and tip the antenna over when not in use — that way, the petunia police won't notice.

The common quarter-wave vertical requires lots of radials if ground-mounted and a few resonant radials if mounted above ground.   The radials make up the other half of a dipole.

Radials reduce ground losses and increase signal strength.  You need a minimum of 16 radials under a ground-mounted vertical.  Each radial is one-quarter wave long on your lowest frequency.  The exact length is not critical, but more radials are better.  Commercial radio stations use hundreds, although the returns diminish after 16.

There are several ways to bury radials.  One is to put the wire on the surface and hold it in place with sod staples.  The grass and thatch will cover the radials and staples within a growing season.  To bury the radials, rent a lawn edger or cut a slit with a spade.  The radials do not have to be deep, just enough to get them out of the way.

Elevated radials don't require as many.  Cut them to one-quarter wavelength with at least two on each band.  They can lie on the roof but keep them from blowing around.

Some verticals require no radials at all. They are designed as vertical dipoles or have matching networks to simulate radials. No-radial verticals tend to be much more expensive, but they do make life easier.

## ANTENNA CHALLENGED?

Don't let an apartment or a community that prohibits outdoor antennas keep you off the air.

I have a friend who loads up his rain gutters. Some hams put an antenna in the attic, but having it that close to electric, telephone, and cable lines is asking for RFI trouble. The antenna will induce current in those wires. It might work with low power, but 100 watts could be a problem.

One solution is a short single-band vertical called a Hamstick. These antennas are 5 to 8 feet long and designed for mobile use. Various mounts and clamps adapt them to fixed use as well. There is a mount that attaches two Hamsticks in a dipole configuration, eliminating the need for a ground counterpoise. Hamsticks are available for $20-$30.

I operated off a condo balcony using a similar short vertical clamped to the railing. The railing acted as the ground counterpoise (radials). It was probably not ideal but worked lots of stations, including DX. When done, release the clamp and bring the antenna inside out of sight to the clipboard-toting-condo-rule-enforcement brigade.

**End Fed** Another solution is an end-fed half-wave or random length. The proper wire length operates on multiple bands. From a mid-rise building, feed the antenna at the top, pulling the bottom as far from the building as you can. From the ground, toss one wire over a tree limb. The end-fed design has become very popular with portable operators. A transformer at the feed point converts the antenna's high impedance to match the coax.

**Portable**  Consider an antenna you can put up and take down easily as used for portable operation. There are many such solutions, most based around a twenty-five to thirty-foot light-weight collapsible fiberglass pole. Run wire up the pole and throw out some radials for a vertical or use the pole as the support for a dipole, Inverted Vee, or end-fed.

Above all, work with what you can and hone your skills. Finesse is more important than power.
I repeat:  **The antenna you have up will work a lot of stations. The perfect antenna you are looking for, but still haven't installed, won't work anyone.**

This Lithuanian station's antenna may be for the birds, but that antenna has the same gain as a 400-watt amplifier.

# REMOTE OPERATION

Advances in Internet technology and high speed connections make remote control stations possible. Communicate using radio equipment manipulated from afar. Audio and control commands pass over the Internet to the remote radio. Remote control stations are a solution for hams crippled by homeowner association rules or other operating constraints. It also might be fun to operate a station in another part of the world and experience different propagation patterns.

Not long ago, I spoke with an amateur using a callsign in the Azores. I recognized his voice, so I asked, "Hey Martti, how's the weather in the Azores?" Martti came back, "I don't know. It is cold and snowy here. I am sitting in my living room back home in Finland." Martti was operating his remote station in the Azores over the Internet. That's a pretty amazing bit of technology, considering the operator was thousands of miles away from the equipment. The control point was in Finland. He was twiddling the knobs from afar.

## REMOTE OPERATION

Here's the quote off the back of Martti's card:
*The technology is here to overcome noise and restrictive antenna ordinances and help you set up a shared station at a reasonable cost with your friends. Or, you can even operate from distant lands without being there.*

*You have now contacted CU2KG Remote in the Azores, with the station operated through the Internet from Finland. You have contacted a highly advanced station employing the latest technology and its people. Just come on board for the thrill of being another remote station, using the Internet as your link.*

There is at least one group offering subscription-service remote-control access. The stations feature amplifiers and monster antenna systems. By the time you read this, there may be several more. Pay a subscription fee and a per-minute fee to control the station from your computer. You may not need to own radio equipment at all!

When operating a remote station outside your country, follow the rules for that country. The transmitter and antenna location determines the operating rules. If the country has a reciprocal licensing agreement with the US, you can identify as CU/K4IA, for example. Alternatively, you could obtain a license in that country. Martti, OH3BH, used his Azores callsign, CU2KG.

Remote operation is an evolving field. Watch for fresh developments in equipment and operating rules.

# MOBILE OPERATION

I've worked mobile HF for years and communicated all over the world using both phone and CW. As a bonus, you can connect your home station to your car-mounted antenna and defeat no-antenna regulations. Just don't forget to disconnect before driving off to work the next day!

Several small HF transceivers designed for home or mobile operation have a detachable faceplate. Only that small control head goes into the cabin. Mount the transceiver in the trunk or under a seat. Secure the transceiver well, so it does not become a flying missile during a car accident or slide around and create a distraction. It is a bad idea to lay a radio on the passenger seat.

Here is an HF/VHF/UHF transceiver in the trunk.

Here is a bracket in the spare tire well securing the radio's mobile mount. Heavy-duty double-sided tape supports the bottom, so it doesn't clatter against the wheel well.

**MOBILE OPERATION**

Run the power cable directly to the battery to reduce electrical noise and provide maximum power. A 100-watt rig draws about 18 amps on transmit, which is beyond the capacity of a cigarette lighter socket. In my case, the battery was also in the trunk, making the connection easy. Fuse both the positive and the negative side of the power cable for maximum protection against overloads and fire.

The wires leading to the top are the microphone, speaker, and control cable. They went under the trunk floor matn. An Internet search found several videos with complete instructions on how to get the back seat out of the way. Route the cables to the front by tucking them under the door trim for a neat and clean install.

The control head mounted on the dashboard to the right of the steering wheel.

The control head attaches with clear Scotch brand "Permanent Double-sided Clear Mounting Tape." It is very sticky but removes leaving no residue.

The next component was the speaker. I was at a loss until I looked in the rear-view mirror and noticed the child-seat restraint on the rear deck in the middle of the back seat. Two zip-ties hold the speaker bracket securely to the restraint clip. I used large zip-ties around a headrest in another install. The wire goes behind the seat to the trunk, where it plugs into the radio.

The speaker looks factory installed.

Before beginning an installation, sit in the car and look for possibilities. I would never have thought of the speaker mount until it jumped out at me.

The microphone holder slipped under the console trim just in front of the driver's seat. Double-sided tape on the bracket under the console keeps it in place.

Mobile antennas are one-quarter wave verticals. They may not be a full quarter-wave long, but an inserted coil electrically lengthens the antenna.

## MOBILE OPERATION

The car acts as the "other" half of the antenna, like radials. This requires a good connection to the chassis. I use an antenna mount that attaches to the trunk lid. Many trunk hinges ride on fiberglass bearings, so there is no bond to the rest of the car. A strap across the hinge connecting the lid and car body will correct that problem.

The antenna pictured below is VHF/UHF. The easiest mobile antenna for HF is a simple one-band Hamstick, mentioned earlier. You may need to use fishing line to guy the hamstick. Attach it from the antenna to the rear seat overhead hand grip.

Seventeen meters is a good band for mobile HF. It is a "gentlemen's band" with good propagation, and the short antenna is reasonably efficient at that frequency.

I call this "the cat's meow."

If you want to get fancy with multi-band HF antennas, there are several to choose. The "screwdriver" types got their name from an electric screwdriver motor that turns a coil to tune the antenna.

The smaller screwdriver versions cover 40-2 meters. The larger go down to 80 meters. The big boys are about 10 feet tall and need a very sturdy mounting system attached to the bumper or the car frame. They would never do on a trunk lip mount.

Dogsled mobile(?) from Greenland.

# INSURANCE

You've spent good money on your amateur station, so think about protecting it. I have lost equipment to a nearby lightning strike and from a power surge. The power surge destroyed a $4,500 amplifier. There was no physical evidence of charred parts or odors. The manufacturer would not attempt a repair because the surge could have compromised parts prone to later failure. I turned in my insurance claim and got paid in full.

A homeowner's, renter's, or automobile insurance policy is not sufficient. Discuss this with your insurance agent, but here are some problems with a typical policy:

- It limits coverage to specifically enumerated risks. Just your luck, something will happen, not on the list.

- It may limit coverage to a specific dollar amount unless you register and pay a fee for additional coverage. That may require you to pay for an appraisal of the equipment's value.

- The policy may have a high deductible. A $1,000 deductible for a $1,000 radio is like having no insurance.

- It may not cover replacement costs and only pay based on a depreciated value.

- The policy may not insure equipment in your car unless it is factory installed.

- It may not cover equipment stolen from an unlocked car or where there is no evidence of forcible entry.

- The policy may not cover damage you do, like dropping a radio or spilling a drink.

- You can't get mechanical breakdown coverage from a homeowner or automobile policy.

Two companies offer insurance specifically tailored for amateur radio stations. They take slightly different approaches to coverage.

The ARRL program requires you to be an ARRL member, list <u>all</u> the equipment you want to insure, and pay a premium of $1.40 per hundred of replacement value on <u>all</u> the equipment. If an item is not listed, it is not insured. There is a $50 replacement and $25 repair deductible. "The Plan does not insure against the usual and customary exclusions such as loss or damage by mechanical or structural breakdown or failure, dishonest acts, wear and tear or damage occasioned by repairing or tuning."

Ham Radio Insurance Associates offers a simplified application, and you do not have to be an ARRL member. You buy an amount of insurance and do not need to list your equipment unless you want more than $5,000 in coverage. You might have $10,000 worth of equipment at home, office, and in your car, but it is highly unlikely all would be lost at one time.

HRIA coverage is $1.50 per hundred of replacement value, but you only buy what you need, so the cost may be less than the ARRL policy. $5,000 coverage, with a $50 per claim deductible, is $75. Optional mechanical breakdown coverage is an additional $25. The mechanical breakdown coverage is limited to radios under five years old and has a $250 deductible. Damage from wear and tear, repairing, or tuning is not covered under either coverage.

Check coverage and costs as they may change by the time you read this.

# OPERATING TIPS AND STRATEGIES

Some hams get through by the brute force of their station's equipment. The rest of us must rely on finesse, tricks, and skills. Finesse can grab the bone from under the big dog's nose. Let's talk about some skills to develop and apply.

## TRAIN YOUR EARS

You've probably heard the advice to listen, listen again, and listen some more. To do that well, first, train your ears.

My first receiver was a military surplus ARC5. I didn't need to tune the dial because the receiver was so broad, it seemed I could hear the whole band at

JUN • 65 •
The ARC5 receiver is on the left.

one time. More than a dozen CW signals on different frequencies came through as distinct tones. I learned an essential skill. Concentrate on one signal and block out the rest.

Concentrate on one signal. Ignore everything else and don't be distracted by other signals, even if they are stronger. Don't allow your concentration to jump around, or you won't copy anyone.

It's a little harder on SSB, but the principle is the same. Concentrate on one signal, and mentally tune out the rest.

## ZERO BEAT

Assuming you haven't engaged the RIT or XIT[26] control, a transceiver's transmit and receive frequencies are the same. Zero-beating sets the receiver frequency to the same frequency as the other station. If you don't tune carefully, you will be off his frequency. He may not hear you or understand well enough to answer. Pay attention, and don't be casual about tuning.

On CW, train to hear one pitch. Pick a comfortable sidetone pitch of, say, 650 Hz, and learn to recognize it. Adjusting the receiver to a tone of 650 Hz will be zero-beat and transmit on his frequency.

Another method is to match the tone heard in the receiver with the side-tone of the transmitter. There is often a "spot" button on the transceiver that will inject the tone. Tune the VFO to match the received signal and the side-tone.

On SSB, tune the signal such that the voice sounds natural. You probably haven't heard the person in real

---

[26] Receiver (RIT) or transmitter (XIT) incremental tuning.

life, so how do you know what is natural?  I tend to tune an SSB signal too high in pitch.  That means he will hear me too low when I call.  I have learned to compensate by adjusting the pitch of his voice down.

Another tip is to look at the frequency.  People tend to call CQ and stay on zeroed frequencies like 7.257.000 MHz.   A receiver tuned to 7.257.078 MHz is probably off frequency.  This trick assumes both your transceivers are well-calibrated.  Maybe he is on 7.255.078 MHz, but it is worth a listen to see if the voice sounds more natural on the zeroed frequency.

## LISTENING

Listening is an important skill to cultivate.  Don't jump on and start calling CQ before listening.

Listening can be very productive because:

- You hear when and where the band is open.
- You may hear another station calling CQ.
- You may discover a DX station before everyone else, giving a chance to work him before the horde descends.
- It's fun.

Tune slowly and listen to every signal to identify the participants.  In time, you will develop a sixth sense to differentiate the interesting from the mundane.  Curmudgeons discussing their gall bladder operations don't interest me.

The quality of the signal can tell a lot.  For example, auroral flutter is a dead giveaway for over-the-pole propagation to Asiatic Russia, India, and the Far East.  Spoken accents may reveal the operator's native language, if not his location.  Raspy-sounding CW may indicate a multi-path DX station or a power supply issue.  Power isn't too clean in many parts of the world, or the station may use an old power supply with leaky capacitors.

I prefer to tune in a direction, so the tone of the received station starts high and descends. It is easier to hear the signal coming. Tuning from below, I can't hear anything until a low rumble ascends to become understandable.

# TAIL ENDING

"Tail ending" is calling at the end of an existing conversation. There is a right way to do this. The wrong way is interrupting.

An interrupter will break in before the conversation is over. One station may say "73," but the other side still has a "final" to deliver. Calling early interferes with the final, irking the listener you are trying to contact. Don't interrupt. Let the conversation finish.

Wait until the station says, "QRZ,[27] "take care," or "good-bye." Wait for something that would indicate the final is acknowledged, and the conversation finished. Then, you can call without being an interrupter. That is the proper way to tail end.

# USE YOUR PASSBAND TUNING

A 2.1 kHz filter is narrow for conversational SSB. Extreme circumstances might call for a 1.8 kHz filter. Those are the standard narrow crystal filter widths.

DSP filtering is not limited to one or two bandwidths, like crystal filters. Adjust the bandwidth and crank it down. When the signal becomes unreadable, adjust the passband. The passband control moves the filter's

---

[27] "Who is calling me?" It implies someone has called. This is often misused as a kind of CQ.

emphasis in relation to the total signal. The default radio passband may not be optimal for a particular signal. Move the passband to the portion of the signal required to understand. That is passband tuning.

I was operating in the Virginia QSO party[28] on 40-meter SSB when the QRM made copy very difficult. I narrowed the DSP below 1.8 kHz and slewed the passband to copy.

The same principle works with a CW signal if there is QRM nearby. Experiment with passband tuning to get away from offending QRM and increase readability.

# REDUCE YOUR RF GAIN

Reducing RF gain seems counter-intuitive. More gain means louder signals, right? Modern receivers are very sensitive; sometimes, too much so. Good copy is all about the signal-to-noise ratio (SNR). You want to hear the signal, not more noise. Turning up the RF gain increases the signal, but also increases the noise level. The signal-to-noise ratio is the same, but the background noise is fatiguing and distracting.

Turning up the RF gain can be detrimental. Receivers have a limited range, the maximum difference between a weak and strong signal, and a limited upper limit or maximum signal. Think of it this way; if you raise the floor in a room, you might bump your head on the ceiling.

Loud signals can exceed the receiver's ceiling or range. This introduces distortion by-products and upset AGC circuits, causing them to mute reception.

The proper setting for the RF gain is to just barely hear the background noise and no further. The signal

---

[28] Most states sponsor mini-contests called QSO parties.

will still be there, but the noise is gone. This setting is less fatiguing for the ears and preserves headroom for the dynamic range limitations of any receiver.

If the natural noise level is high, as it usually is on the lower HF bands, switch in the attenuator. The attenuator reduces everything coming into the radio before it reaches the sensitive RF amplifier stage.

A pre-amplifier boosts the signal before it goes to the RF amplifier. Never use the pre-amplifier except on higher-frequency bands (15 meters and up) and only when the background noise is low.

I remember watching over the shoulder on a 75-meter phone position at Field Day. The operator had the pre-amplifier on with the RF gain cranked up to 11. He complained that all the noise made it impossible to copy. I asked politely if I could make some adjustments and the frustrated op agreed. I turned off the pre-amp, reduced the RF gain, and switched in some attenuation. The band cleared up, and signals became readable. If the radio is too noisy, turn down the RF gain and switch in attenuation.

# CW STRATEGIES

Be careful to zero-beat and don't exceed the other operator's speed. You'll be the slowpoke in the beginning, but speed will improve with practice.

If you need your QSO partner to slow down, send "PSE QRS," "Please send slower." Most operators will be courteous and comply. If he doesn't slow down, tell him politely that you can't copy, and hope to see him again soon.
"SRI OM NO COPY  HPE CUAGN SOON
73 WA4TUF DE K4IA SK"

# PHONETICS

Standard phonetics cut through noise and QRM. They are especially important when dealing with different languages. Saying "A" as in "date" to a Hispanic, he sees the letter "E." "E" is the letter "I" to him. No wonder callsigns get busted.[29]

The FCC assigned me KG4CVN when I was re-licensed in 1999. "G," "C" and "V" sound alike, especially in noise or fading conditions. So do "N" and "M". Golf, Charlie, and Victor sound nothing alike. No one can confuse November and Mike.

Standard phonetics are easier for others to understand. It is awkward at first but make it a habit. Learn the phonetic alphabet and it will come naturally. Notice I said, "standard phonetics." Cutesy or non-standard variations only confuse.

Be consistent. If you are Kilo Alpha 4 India Alpha, don't say Kilo America 4 India Alpha. You are using America and Alpha for the same letter, confusing your QSO partner.

Be clear and crisp. Slow down and enunciate. Don't mumble, don't shout and don't stretch out the callsign. Keeeeelllloooo is not easier to understand than Kilo. It can sound like several letters – Kilo, Hotel, or Tango depending on which part of the elongated word he heard. Keee can sound like Kilo. Ell can sound like Hotel. Oooo can sound like Tango.

---

[29] A "busted" callsign is one copied incorrectly.

Here is the word list adopted by the International Telecommunication Union:

| | | |
|---|---|---|
| **A**--Alpha | **J**--Juliett | **S**--Sierra |
| **B**--Bravo | **K**--Kilo | **T**--Tango |
| **C**--Charlie | **L**--Lima | **U**--Uniform |
| **D**--Delta | **M**--Mike | **V**--Victor |
| **E**--Echo | **N**--November | **W**--Whiskey |
| **F**--Foxtrot | **O**--Oscar | **X**--X-ray |
| **G**--Golf | **P**--Papa | **Y**--Yankee |
| **H**--Hotel | **Q**--Quebec | **Z**—Zulu |
| **I**--India | **R**—Romeo | |

# GET A NEW CALLSIGN

The FCC assigned randomly your Technician Class callsign.  It is probably a confusing mouthful.  Get a new callsign that is shorter and easier to say.  I attribute a lot of my operating success to my call, K4IA.  It sure is easier than KG4CVN, the one assigned when I was re-licensed in 1999.

Avoid tongue-twisters but also think about how a callsign sounds in Morse code.  You might operate CW one day.  For example, the suffix "JYQ" badly combines the longest letters.  See the QSL cards on the next page for some more examples.  5H3/K8Lee is 12 dits in a row.

The FCC vanity callsign program allows you to pick from a list of available calls.  An internet search will tell about it and how to apply.  The FCC charges $35 but it may be the best $35 you spend.

*An excursion during the work in South Africa*

*Another DX operation by JH4RHF, Sep-Oct 2005*

UX5UO print

Pity poor Jun, a Japanese operator using his home call in Lesotho. That home call, JH4RHF was tough enough, but adding the 7P8 to the front made it a real effort to copy. I worked him on CW, and it sounded like the callsign would never end.

## DAR es SALAAM, TANZANIA

# 5H3/K8LEE

EX: W9OEH, CYØAA, VP2MNQ

Cfm QSO w/ WA4TUF

| | MHz | RST | MODE 2-WAY | QSL |
|---|---|---|---|---|
| 2WAY:CW    MHz:14.025 | | | | PSE |
| ON:21-OCT-99   UTC:03:22   RST:599 | | | | TNX |

TNX QSO/QSL 5H3/K8LEE

**WAYNE McKENZIE**
24815 Joy Lynn Road
Lawrenceburg, IN
47025    U.S.A.

The QSL MAN® - W4MPY

Here's another one that is tough to copy. On CW, 5H3 (Tanzania) is 12 dits in a row!

# ETIQUETTE, DECORUM AND PROPRIETY

*Etiquette, decorum, propriety imply observance of the formal requirements governing behavior in polite society. Etiquette refers to conventional forms and usages: the rules of etiquette. Decorum suggests dignity and a sense of what is becoming or appropriate for a person of good breeding: a fine sense of decorum. Propriety (usually plural) implies established conventions of morals and good taste: She never fails to observe the proprieties.*[30]

Etiquette, conventional forms, and usage, falls short of decorum, a sense of what is appropriate, and propriety, conventions of morals and good taste. I am going to lump them all together into "do's and don'ts" of operating.

So here are my Proverbs:

Do listen before calling.

Don't call CQ unless sure the frequency is clear.

Don't call until you know what to expect.

Do listen more than transmit.

Don't use "sharp elbows" by continuously calling and stomping over the competition (and yourself).

Don't call out of turn as when the other station calls someone else.

Do wait before calling again.

Don't QRM your QSO by calling too soon and covering his reply.

Do wait for your turn.

Don't tail end improperly. Wait until sure the previous conversation has ended.

---

[30] Dictionary.com

## ETIQUETTE, DECORUM AND PROPRIETY

Don't call when a station calls someone else.

Do call with your full callsign.

Don't use partial calls. Partial calls are confusing and require an additional exchange to complete a contact.

Do use standard phonetics.

Don't use cutesy or inconsistent phonetics.

Do pay attention to the operator's instructions.

Don't call if he is working another area or number.

Do tune up into a dummy load or a clear frequency.

Don't tune up on an existing conversation.

Do ignore smart-alecks, jokesters, QRMers, and malcontents.

Don't taunt them. It only encourages more bad behavior. Leave them alone; they will get bored and move on.

Do check your transceiver before calling.

Make sure you are on a frequency authorized for your license class and watch that your sidebands are in the band.

Do maintain a tolerant world-view. Talk about ham Radio.

Don't ask questions like, "When are you going to get rid of that bum governor?"

Don't drag up religion and politics.

Do identify with your callsign.

Don't hide behind anonymity to make rude remarks. If you are not willing to put your callsign on something, you shouldn't say it.

Don't make unidentified illegal transmissions.

# DX OPERATING

If you hear a station calling, "CQ DX," he is looking for a contact outside his country.  The term "DX" means "distance" but also refers to foreign contacts.

International stations are not rare, and the magic of radio means your signal travels a long way.  There are many easy to contact entities in Europe, Central and South America, the Caribbean, and Japan with thousands of operators.

Don't be intimidated by DX.  Many of those operators are as green as you.  They may have minimal stations in tiny yards or apartments with compromise antennas, low power, and antique or homebrew equipment.  The  signal at his end might be louder than the signal you hear.  Don't assume because a station is weak, you won't be able to work him.

DX stations look for their Worked All States (WAS) award.  There are quite a few county-hunters as well.  I remember working an Englishman while driving through the Virginia countryside.  It thrilled him to get several new counties.

DX operators might want to practice their English.  Despite what you might hear from the mean-scream media, the USA is looked upon with favor by most of the world.  The point is, they are as eager to make contact as you, so get over "mic fright" or "key clutch" and make the call.

Most DX QSOs are short.  Don't worry about having nothing to say.  In a DX QSO you will exchange a signal report, QTH, and name.  I call that the QSO trinity.

Proper etiquette says the station that had the frequency controls the conversation.  Listen to what he is doing.  If it is just a signal report, do that.  If he is more talkative, give him the QSO trinity.  If he wants more, he will offer information about his station, the weather, how long he's been a ham, how he QSLs, etc.  Then, reply with yours. If it is just the QSO trinity, don't say more.  It could be his English is very limited, or maybe he is just trying to make contacts. If he doesn't offer more, ask for a QSL card, thank him for the contact, and get out of the way.

TOKYO JAPAN    JCC #1001
                ZONE 25

TNX FB
QSO.

**JK1UVP**

*Kazuo  Mori*

2-2-10, Higashimatsumoto, Edogawa-ku,
Tokyo 133-0071 Japan [GL:PM95]

Don't be intimidated.  Amateur radio and sumo wrestling are very popular in Japan.

If you hear a DX station calling CQ, QRZ or ending a conversation (be sure he is really done), you would give a short call to him like this: "JK1UVP this is (or "de" on CW) Kilo 4 India Alpha, Kilo 4 India Alpha." More often, the station calling the DX will just say his callsign, "Kilo 4 India Alpha."  Listen to the previous conversation to learn what to expect.

I always use phonetics on my call because letters are hard to decipher amid noise and fading.  Language is also a barrier.  See the section on Phonetics in the Operating Tips and Strategies chapter.

Don't use phonetics on his call. I am sure he knows his call. And don't be repetitive saying his call several times or your call more than twice. He knows you are calling him because you are zero-beat on his frequency (hopefully). Who else would you be calling? If he doesn't copy your call, he will say, "QRZ?" or "Who is the K4?" and you try again.

Speak clearly and enunciate. Keep it simple. Avoid idioms, slang, complicated words, and tortured sentence structures. If you tell a non-English-speaking person the band is "dead as a doornail," he probably won't have a clue what you mean. He won't know what a "bodacious" signal is either.

Digital modes such as CW, PSK, FT8, and RTTY have a huge advantage over SSB in that they can copy much weaker signals. Add CW and digital to your operating modes for maximum DX.

Rare DX is the entity not often heard. Maybe there are only a few hams living there, or the entity gets visited sporadically, or propagation to that part of the world is difficult. Usually, all three strikes are against you. You can reasonably expect to work them, but it is not going to be "casual."

The rare DX station may be a laid-back operator in an unusual place. He may be no more experienced than you are. His equipment might be no better than yours. He may hate pileups and think sending QSL cards is an expensive chore. He has his license to communicate with the outside world but is not your DX puppy. These laid-back operators are likely to respond to a CQ. They can have a contact without fighting a pileup and then be gone. I am surprised many times by a call I never expected.

Rare DX QSOs are usually nothing more than a signal report. Once again, the station that had the frequency dictates the protocol.  Listen for the operator's pattern.  If the DX is tearing through a crowd like an octopus on crack, the QSO Trinity of Report, QTH and Name shortens  to just a signal report, and it is always 59 on SSB and 5NN on CW.  5NN is short for 599.

Call the DX by giving your full callsign once.  Notice I said <u>full</u> callsign.  Don't use his callsign at all.  Listen for 5-7 seconds, and if he hasn't come back to anyone, throw out your full callsign again.  If he calls you, say 59/5NN and maybe "73' or "Thanks" "TU;" not one more word unless the DX initiates it.

Rare DX may require you to work "split."  When a rare DX station appears, it doesn't take long for the world to find out.  Suddenly, there can be hundreds of stations calling.

"Working split" means the DX is transmitting on one frequency but listening on another.  Transmit where he is listening and receive where he is transmitting.  You need to learn how to work split to crack into the rare DX level.  I explain more in my DX book.

# DX CODE OF CONDUCT

A group of prominent DXers published a set of operating guidelines called the DX Code of Conduct. The rules make sense for domestic contacts as well. Considerate operating is better and more enjoyable for everyone.

1.    *I will listen, and listen and then listen again before calling.*

2.    *I will only call if I can copy the DX station properly.*

3.    *I will not trust the Internet spotting network and will be sure of the DX station's callsign before calling.*

4.    *I will not interfere with the DX station or anyone calling, and I will not tune up on the DX frequency or the QSX (split) spot.*

5.    *I will wait for the DX station to end a contact before calling.*

6.    *I will always send my full callsign.*

7.    *I will call and then listen for a reasonable interval.  I will not call continuously.*

8.    *I will not transmit when the DX operator calls another callsign, not mine.*

9.    *I will not transmit when the DX operator queries a callsign not like mine.*

10.    *I will not transmit when the DX operator requests geographic areas other than mine.*

11.    *When the DX operator calls me, I will not repeat my callsign unless I think he got it incorrectly.*

12.    *I will be thankful if I do make a contact.*

13.    *I will respect my fellow hams and conduct myself so as to earn their respect.*

# DX CODE OF CONDUCT

Don't be an elephant, stomping over everyone attempting to make contact.

United Nations Headquarters Station, New York.
Some DX pileups might need UN peacekeepers.

# PROPAGATION

I can't get into the exact details of possible propagation from your QTH this week.  Conditions change depending on many variables, and there are plenty of sources for current predictions.

Getting started on HF, you'll be happy to make any contacts at all.  Learn about propagation, so you don't waste time calling or searching a dead band.

The general rule is higher frequencies are open during the day and periods of high solar activity.  The lower frequencies open at night.  Twenty meters is in the middle and can be open somewhere 24 hours a day.  Above and below twenty meters, the general rule is more noticeable.

To be more specific:

- 80 meters – nighttime band.

- 40 and 30 meters – mid-range during the daytime (500 miles) and possibly worldwide at night.

- 20 meters – the "go to" long-distance band is usually open to somewhere 24 hours varying in distance with solar flux and time of day.  Can be open from the East Coast to Australia and the Pacific around midnight.

- 17 and 15 meters – 17 acts much like 20 and 15 needs a bit more boost from solar activity.  I think of them as afternoon and evening bands.  They often feature north-south openings to South America and the Caribbean.

- 12 and 10 meters – Afternoon and evenings during periods of solar activity.  When open, a little power will go a long way because the noise level is lower on these higher frequencies. .

## PROPAGATION

The problem with 80 and 40 is they are noisy and you can hear thunderstorms from thousands of miles away. That can be a real distraction in the summer. They also suffer from more man-made noise. The higher frequency, shorter wavelength, bands have less noise, which is a significant advantage for hearing weak signals.

The Maximum Usable Frequency (MUF) is the highest frequency that might support propagation on a given path and time. Listen "up" on the band closest to the MUF to take advantage of the quieter conditions found there.

# PREDICTION PROGRAMS

There are several propagation prediction sources. Bandconditions.com will provide an overview.
The DxLab Suite has a module called PropView that creates a chart of Maximum Usable Frequency and displays band openings.

VoCap.com/prediction can forecast openings from your location to a DX station and generate a prediction chart for various bands and times.

# BEACONS

Real-time beacons come from the NCDXF/IARU[31] Beacon Network. Eighteen beacons broadcast from different parts of the world on a ten-second rotating schedule.

Tune to a beacon frequency and hear the stations as they switch from location to location every 10 seconds. It takes 3 minutes to cycle around the world. When a cycle finishes, you know the band conditions.

---

[31] Northern California DX Foundation / International Amateur Radio Union.

The band is open to the parts of the world where you hear the beacon. It will amaze you at how little difference there is in signal strength as the beacon sends at lower and lower powers. If you can hear them at 100 watts, you will almost always be able to hear at the lower levels as well.

Beacons broadcast in this order from New York, Northern Canada, California, Hawaii, New Zealand, Australia, Japan, Russia, Hong Kong, Sri Lanka, South Africa, Kenya, Israel, Finland, Madeira, Argentina, Peru, and Venezuela.

Each station operates with a simple vertical antenna. The beacon identifies in Morse code, running 100 watts. Then, it sends a dash at 100 watts, 10 watts, 1 watt, and 100 milliwatts (1/10 watt). If you don't know Morse code, the website shows who is transmitting at NCDXF.org/beacon/beacontools.

Beacon frequencies are 14.100, 18.110, 21.150, 24.930, and 28.200 MHz in the CW portions of the bands.

## CALL CQ

Calling "CQ" into a dead band might bring a very surprising contact. If it doesn't, what did you lose? No one is going to laugh at you for calling CQ. And if the band is dead, no one will hear, anyway. What else are you doing? Disturb some electrons. I guarantee the electrons won't complain.

Rather than  listening to static, call CQ a few times and see what happens. Better yet, set up a keyer to call CQ for you. A voice keyer stores and plays back a recorded voice message, and a CW keyer does the same for CW. Either can auto-repeat after an interval, so it keeps going without intervention. The keyer

## PROPAGATION

never tires, and it never gets bored. Mash the button, sit back, and wait for an answer. Just don't walk away and leave it running!

Like a fisherman chumming the water wait to see what comes up to take the bait. Read the paper, balance your checkbook, or watch the ball game while the machine does the work.

Listen and ask if the frequency is in use before calling CQ. Even if the band sounds dead, the frequency may be in use, so ask. Identify first, then ask. Listen a few seconds and ask a second time.

Send "QRL?" on CW. If the frequency is in use, you will hear "C" ("yes") in response.

Don't be a blockhead. Make sure the frequency is clear before you call CQ.

# ANATOMY OF A QSO

## THE WRONG WAY

Now you've learned the rules. It's time to look at an actual QSO. First, we'll consider how Larry the Lid[32] might go about it.

Larry has a big station with lots of power and one of those Windom antennas from Carolina that claims 13dB gain in all directions. Larry likes to be loud and never operates without his amplifier. Across the state, across the country, or around the world, Larry pours on the coal.

It is getting near time for Larry and his buddies to hold their daily gabfest. He can't wait to hear the latest on how Moe's cat reaction to the new mange medicine. Curley's gripes about city hall are enough to make anyone join the militia. Larry is eager to share his experience at BurgerDoodle. He set the counter girl straight when she messed up his double-bacon-diablo-big-bull order. "Yeah, man, I had her in tears. I can't wait to tell the guys all about it."

Larry turns on his radio and is chagrined to find there is someone on HIS frequency. "Damn, how am I going to shoo him away? Let's start by checking my amplifier; surely, it needs tuning up." After a lengthy key-down, Larry switches back to receive, and his "intruder" is still there.

Larry doesn't understand propagation. The intruder is talking to someone who can't hear Larry. Many times, you cannot hear both sides of another's conversation. Larry's attempt to disrupt has no effect.

---

[32] A "Lid" is a poor operator. Don't be a Lid.

## ANATOMY OF A QSO

Larry gets discouraged and looks at the clock. He has another hour before the boys gather, so he decides to call CQ. What better place than right here on "his" frequency. It never occurs to Larry that he should wait and let them finish. Larry doesn't bother to listen. If he did, he would learn they are wrapping up their QSO, and "his" frequency will be free long before his sked.[33]

Listening is too much of a bother. Larry cranks up and starts calling and calling and calling. "CQ CQ CQ twenty meters CQ CQ CQ twenty meters CQ CQ CQ twenty meters CQ CQ CQ twenty meters this is W5LID,[34] W5LID, W5LID, Wanted For Living In Dreamland. CQ CQ CQ  CQ CQ CQ  CQ CQ CQ  CQ CQ CQ  CQ CQ CQ DOOOOOUBLEUUUUUU FOOOOOOR ELLLLLLLLL IIIIIIIIII DEEEEEE. Calling CQ twenty meters. Over Over"

Silence for two seconds. No reply, so he calls again and again and again repeating "CQ CQ CQ  twenty meters CQ CQ CQ twenty meters CQ CQ CQ  twenty meters CQ CQ CQ twenty meters this is W5LID, W5LID, W5LID, Wanted For Living In Dreamland. CQ CQ CQ  CQ CQ CQ  CQ CQ CQ  CQ CQ CQ  CQ CQ CQ DOOOOOUBLEUUUUUU FOOOOOOR ELLLLLLLLL IIIIIIIIII DEEEEEE. Calling CQ twenty meters. Over Over"

Somewhere in there, the intruder's conversation has ended but, since Larry is transmitting, he doesn't hear. Larry keeps pounding away. Larry is unaware that his aggressive approach has made him unpopular, and people avoid talking to him.

---

[33] A "sked" is a scheduled meeting.

[34] I don't think there is anyone with that callsign, but if so, you have my apologies and sympathy.

Finally, a station answers Larry's CQ. "W5LID, this is Kilo 4 India Alpha Kilo 4 India Alpha, how copy?" Larry thinks, "He's only S9. The dumbazz must not have an amplifier."

"Hey K4IA, you're 59 in East Scuffboot Texas. Handle is Larry. That's not Barry, not Jerry, not Gary, not Perry, not Terry, not Harry, and not Kerry. I'm waiting for Moe and Curley to come on our frequency. We've been using this spot for years. I've only got time for a short one, and then you need to move right away."

I am not sure how to respond to that. What was his name? Why did he bother to call CQ? He doesn't want to talk. He was tying up the frequency denying it to others. I should just spin the dial and look for another, more interesting contact. I decided to be polite and respond, "Sorry to hear you're so pressed for time. You're 20 over in Fredericksburg, Virginia. Name is Buck. Enjoy your frequency. 73. K4IA clear."

Larry smiles to himself. "I guess I fixed that guy." Moe and Curley will be on in a few minutes. Wait until I tell them how I saved our frequency." To kill some more time, Larry loads a tape of his rambling diatribe against the FCC. Then, he goes upstairs to refill his bourbon while the recording plays out over the air.

This scenario is not far-fetched at all. There are a few bullies and discourteous amateur radio operators. You will hear Larry-the-Lids hogging a frequency, playing pre-recorded rants, discussing inappropriate subjects, and running off intruders. The best strategy is to ignore them. Don't engage in arguments and don't lecture them on proper behavior. They have heard it before and don't care. Your transceiver has a frequency dial. Use it and move somewhere else.

Eventually, Larry will grow frustrated and drop out of amateur radio. He will complain that it is not fun any longer. Larry could also attract the attention of the FCC and draw a hefty fine or forfeit his license. Too bad Larry ruined this great hobby for himself and others.

# THE RIGHT WAY

Always check your frequency before transmitting. Be sure you are within the limits of your license class. You will hear Extra Class licensees in the lower portion of the band. Foreign stations on 40 meters have privileges where US phone stations can't operate. Don't assume because there are voices you can transmit there as well.

What is the right way to start and sustain a conversation? We would handle DX differently, so for the moment, assume a domestic QSO.

I don't recommend you call CQ until you get familiar and comfortable conversing on the radio. Listen and find someone else calling CQ. Why?

Ham etiquette says the person who called CQ sets the tone and pace of the conversation. Admit it; you don't know what you are doing. Mr. CQ goes first. He gives a signal report, QTH, and name. You give him yours. He might describe his equipment. You tell him about yours. Let Mr. CQ lead. Don't panic about, "What do I say next?" Follow the other guy's prompt and respond to him. Think of a question to keep the conversation going.

Keep transmissions short. Regularly spoken conversation is one or two sentences, and then the other person speaks or interrupts. You can't hear the other side with the PTT switch mashed. You could ramble on forever, and no one can interrupt.

Repeaters time out. There is no such protection on HF. Long discourses are tough to follow. Don't be a hog. The other person should not have to take notes, so he remembers your 14 points before responding.

Think about that while he is speaking. Are you losing your place? Is he going on too long? Pace yourself accordingly and learn from the experience. Exchanges are more interesting if they proceed like a regular conversation. The pattern should be brief two-ways and not extended monologs.

Answer a CQ by giving his callsign once and yours twice. Use phonetics with your callsign. "WA4TUF this is Kilo 4 India Alpha, Kilo 4 India Alpha." You might add, "How copy?" or "Over."

One established, a simple, "What do you think, John?" is enough to let the other guy know it is his turn. A conversation does not require identifying on every "over." Identify every ten minutes and at the end of your contact, as the rules require.

When ready to lead an HF conversation, call CQ after ensuring you're in band and the frequency is free. Sometimes, you can't hear both sides talking, so identify and ask, "This is K4IA. Is the frequency in use?" Listen for a few seconds and ask again.

Hearing no one, call CQ. "CQ CQ CQ this is K4IA, Kilo 4 India Alpha, Kilo 4 India Alpha, calling CQ and listening."

I prefer several brief calls to one long one. Listeners will tire of a long CQ such as Larry's and spin the dial. If a short CQ doesn't get an answer, call again after at least five seconds.

When answered, acknowledge the station. "WA4TUF, K4IA returning. Thanks for the call. You're 58 in

## ANATOMY OF A QSO

Fredericksburg, Virginia. Name here is Buck, Bravo Uniform Charlie Kilo. How copy?"

Say his call once. He knows his call, and presumably will recognize it. I don't need to repeat my call. He knows it because he called me. I might repeat the signal report if conditions are poor but usually once is enough. I don't spell out Fredericksburg phonetically. It is too long. If the other guy doesn't get it the first time, he can ask. I spell out my name, maybe twice if conditions are poor or the other guy sounds weak.

That initial exchange is the classic QSO trinity, report, QTH and Name. Where it goes from there is up to you. Next time, I might share more information with him, where Fredericksburg is, my equipment, my weather, hobbies, job, how long I've been a ham, or my favorite ham radio activities. That would take several transmissions.

So far, I have only talked about myself. To be a good conversationalist, ask the other guy about things that interest him. "My rig is a WizBang 4000 Pro. What's yours?" "I like to garden. What do you do when you're not on the radio?" Find a common ground or learn something new.

You run out of things to say, eventually. How to end the conversation? You could take the coward's way out and claim the XYL[35] is calling, the dog is barking, or the phone rang. There is no need for subterfuge. It is acceptable to say simply, "Thanks for the contact, John. I think I'll try to scare up a few more before I go to bed. Hope to hear you again soon. 73 and best DX. WA4TUF this is K4IA clear on your final." Adding, "Clear on your final'" lets listeners know you are not entirely done and discourages tail-ending interrupters.

---

[35] XYL is hamspeak for wife.

Ending a radio conversation is not nearly as awkward as walking away from a cocktail party windbag.

You will meet some fascinating people. Take "Tex" here. Tex, W5 Big, Quick, Ugly was 101 when I talked to him on my commute home. He was quite a character.

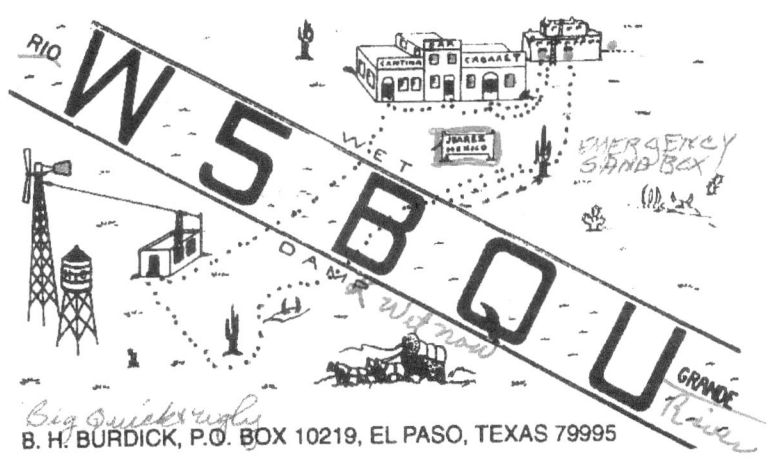

At the time we talked, Tex was reputed to be America's oldest active ham. He is a Silent Key now.

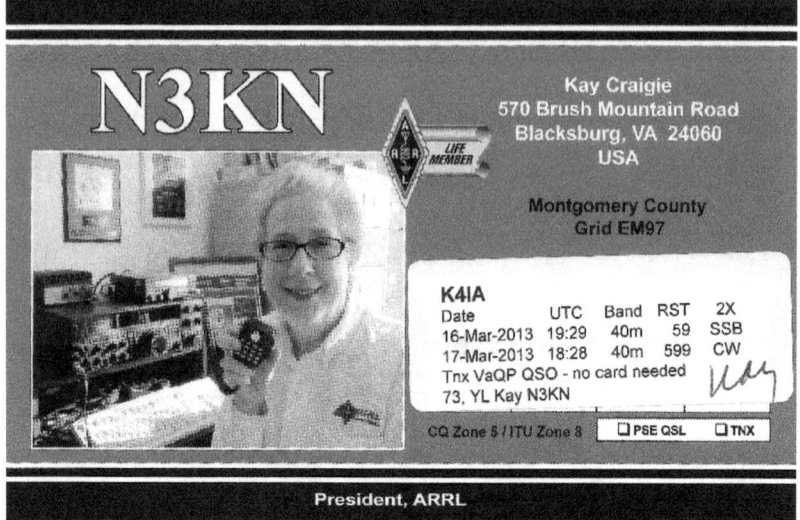

Kay was ARRL President when we met on the air.

Caveman portable.

# CONFIRMING YOUR CONTACTS

Confirmation of your contact is required for it to count toward most awards. The QSL card is one way. Logbook of the World is another. *CQ Magazine* recognizes eQSL, an online log, for some awards.

Confirmation from each of the 50 states earns the Worked All States award from ARRL. Even if you aren't a certificate chaser, collecting QSL cards is a fun part of the hobby.

## QSL CARDS

The old saying goes, "A QSL card is the final courtesy of a QSO." QSLs have been around since the beginning of ham Radio. I love collecting them, and some of my favorites are on the pages of this book. I have shoeboxes full and it is fun to reminisce while looking at cards from past contacts. The foreign cards from exotic lands with beautiful pictures, strange handwriting, and unique designs are a special treat.

Let's start with your QSL. You can go cheap, but printing full-color QSL cards is not expensive. Try something interesting enough to elicit a response. A plain card is sufficient for confirmation, but I like to send something nice, so the other guy feels like I cared enough to deserve his card in return.

I put my card in an envelope for protection, and I will often include a brochure from our local visitor's center touting all the town's historic attractions. Mailing your card as a postcard saves a few pennies, but they look ratty after the post office gets finished with them. I once had a mail carrier ask me if I knew how they came up with the forty-seven-cent stamp price. I said, "No.," and he replied, "Seven cents for delivery and forty cents for storage."

## CONFIRMING YOUR CONTACTS

Unique cards are more interesting to receive, and I have to believe they are more likely to generate a response.

**WAITING FOR AMERICA AND GETTING CHINA INSTEAD!**

Who can resist answering this card from Northern Ireland?

Ron's card is a unique work of art.

Your caricature conveys a personal touch.

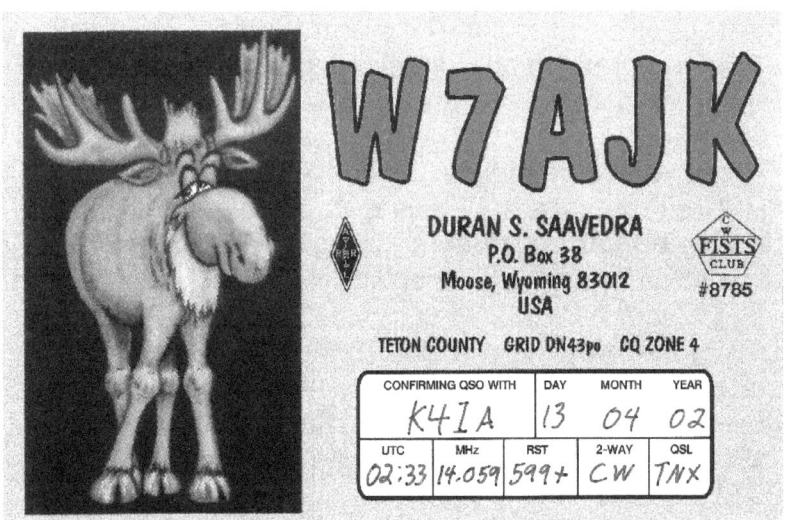

Moose, Wyoming. Who could have guessed?

## CONFIRMING YOUR CONTACTS

This ham lives where Washington crossed the Delaware.

I didn't find self-printing cost-effective or worth my time. I was spending valuable radio time fussing over artwork and printer issues. Many professional QSL printers offer full-color and picture cards. They will do a much better job and be less expensive than you buying cardstock, ink, etc.

Fill out your card legibly. It doesn't help if the reader can't tell the date or details because of sloppy handwriting or ink smears. Logging programs offer the option of printing report labels to stick on the card.

Is it month/day/year or day/month/year? Foreign stations sometimes put the day first. The military and government often put the year first. What date is 11/07/10? Nov 7, 2010, July 11, 2010, or July 10, 2011?

Use English or use Roman numerals for the month. Common abbreviations are acceptable. Then, write out the four-digit year: Nov 7, 2019, XI/7/2019. Anything done to make the responder's job easier will pay off with a higher rate of QSLs confirmed.

A computer logging program will transcribe the contact information correctly.  If you are hand logging, check and double-check the data.

Here is the card I have used.  The picture came from Field Day, and the cigar helped control mosquitoes.

Fredericksburg, Virginia USA
Spotsylvania County

## QSL ETIQUETTE

Once you have selected and completed a card, you need to get it to the other station and hope for a return confirmation.  There was a time when a card sent to another US station almost always got a return.  Sadly, it seems the art of paper QSLing is dying.

My card is free for anyone who works me.  I am glad you took the time to put me in the log and happy to QSL.  I wish everyone would do the same.

Include a self-addressed-stamped-envelope (SASE) to improve the chances of a reply.  To maximize your return and minimize your cost, ask the other operator, "Hey Joe, I am new at this and would like to get your QSL card.  Do you QSL?"  Then, he will tell you what to expect.

**CONFIRMING YOUR CONTACTS**

QSLing DX stations is a topic unto itself, covered in my DX book. Here's the short version.

You can send cards directly to the DX with address information from QRZ.com or other callbook services. If you QSL direct, the DX expects a self-addressed envelope. Use the self-sealing kind, so he doesn't have to lick hundreds of returns.

Don't send American postage stamps because they aren't good for postage from a foreign country. Enclose a few "Green Stamps" for the return postage. Green Stamps are dollar bills. The cost of printing and mailing QSL cards can be staggering for a foreign ham. It takes at $2 to $3 for return postage from another country. Add $1.50 for your postage, and you can see that QSLing DX directly is expensive.

# QSL VIA A MANAGER

Some DX uses a QSL manager. The DX sends his log to a kind-hearted volunteer who handles the QSL duties for him. Send your card and a self-addressed-envelope (self-sealing) to the QSL manager. In the course of time, you'll receive a confirmation. If you're fortunate enough to be dealing with a manager in the US, put a US stamp on the return envelope. Otherwise, enclose green stamps.

# QSL VIA BUREAU (BURO)

An alternative for direct QSL to a DX station is via the QSL Bureau system. Here's how that works:

ARRL members can bundle up cards, sorted by country, and send them to the ARRL Outgoing QSL Bureau. ARRL charges based on weight to collect the cards and forward them, in bulk with others, to the foreign country's QSL bureau service. The foreign service distributes the cards to individual members.

Not all countries take part.  That information, along with additional details, are on the Outgoing QSL Bureau page of the ARRL website.

For incoming cards, find the person who manages your callsign on the ARRL incoming QSL bureau website.  Send either stamped envelopes, cash, or a check so he can purchase envelopes and postage.  Contact the bureau manager to see what he prefers.

Incoming cards get bundled together as received and sent when an envelope is full.  It is like Christmas when that envelope arrives, full of cards from all over the world.  ARRL membership is not required to receive cards.

This system is economical, but the round trip can be slow.  I have had some take years.  I think my record is eleven years.  Lots of DX uses the bureau system, and if you are active, you can count on a steady stream of beautiful and exotic QSL cards.

On the next page are two cards from A to Z.  There are 341 recognized entities as of this writing, so you've got lots of possibilities.

## CONFIRMING YOUR CONTACTS

Gaborone BOTSWANA

# A22 / JA4ATV

A22 is the prefix for Botswana.  A Japanese ham was operating there.  Hence the callsign A22/JA4ATV.

## Greetings from the south of Africa

Artur Makhtsiev, ex UA6JFF, RA6JF

### ☑ ZS6BQI   ☐ 7P8BA

Joburg, Republic of South Africa          Maseru, Lesotho

| Radio | DD/MM/YY | UTC | MHz | Two way | RST |
|-------|----------|-----|-----|---------|-----|
| K 4 I A | 22.12.07 | 1848 | 18 | CW | 549 |
| via: | 10.01.09 | 2101 | 10,1 | CW | 579 |

RV1CC print

☐ PSE QSL TKS ☑ via buro or:   WAZ: 38  ITU: 57

P.O.Box 44698, Linden 2104, South Africa          73!  *Art*

ZS is South Africa.  Art has a license there and in Lesotho.

# LOGBOOK OF THE WORLD (LOTW)

A DXpedition or contest station might make tens of thousands of contacts a year. An active HF operator might make hundreds. The burden of sending paper QSL cards is just too onerous, not to mention expensive. I love and prefer paper QSL cards, but I understand the problem.

ARRL has developed an online verification service called Logbook of the World. The premise is that stations upload their logs, and the system matches up contacts to create an electronic confirmation that can be used for ARRL and some *CQ Magazine* awards.

LOTW is free, but a US amateur must be an ARRL member and pay for award credits. Presently, the application fee is $7.50 plus 12 cents per credit, so DXCC of 100 credits would cost $19.50 plus another $12 for the paper certificate to hang on the wall.

Once successfully registered and authenticated, upload logs from your logging program or by an ADIF[36] file, and LOTW does the work. Most computer logging programs will upload automatically. There is no paper card, but confirmations come rather quickly for little or no cost. Compared to other methods, LOTW is lightning fast and cheap.

LOTW is available for domestic and foreign contacts, but not everybody participates. I upload everything. When someone joins the system later, LOTW will match our contacts automatically.

As I write, there are over 1.7 billion QSO records in the system, and over 391 million QSL records have resulted. I have over 49,000 QSOs uploaded and

---

[36] Amateur Data Interchange Format – a standardized file format for amateur radio programs.

## CONFIRMING YOUR CONTACTS

more than 23,500 confirmations. The money saved on those confirmations is substantial.

An electronic confirmation cannot substitute for a real paper card from somewhere like Kazakhstan.
This is one of my favorites.

Gyuri is QRL (busy) chasing QSL cards.

# COMPUTER LOGGING PROGRAMS

A log of your activities is helpful and not just for when the FCC comes calling. You can't possibly remember your prior contacts, or QSL data, or if you've worked Wyoming on 40 meters or when you changed your antenna.

A paper log is a start, but you can't easily sort or search in a paper log. Data management is what computers do best, and there are several computer logging programs, both paid and free.

I am most familiar with DxLab Suite by AA6YQ, Dave, so that is the one I will describe here. DxLab is free; there is a very active users group and a Wiki. Dave tirelessly advises and provides upgrades. Other popular logging programs offer similar functionality, including N3FJP, Ham Radio Deluxe, and Logger32.

DxLab is not a contest logging program. Contest logging programs are specialized software written to keep tabs on contest exchanges and scoring. They are not intended to track QSLs and awards. Fortunately, contesting programs, such as N1MM+, WriteLog and N3FJP will generate an ADIF file. Import that into DXLab or LOTW to keep your contact information together.

DxLab is a suite of coordinated programs. Entering data in one module populates the others. I will describe the functions but don't have room to be exhaustive. A suite of programs that does as much as DxLab is going to have a learning curve. Don't despair. Visit the DxLab website and see what it offers. This explanation focuses on functionality.

**Launcher** is the gateway into the DxLab system. When Launcher starts, it checks for program and database updates and offers you the option to

## COMPUTER LOGGING PROGRAMS

download and install them. It then starts the various modules that make up the suite.

**Commander** interfaces the radio with the computer through a serial port or USB cable. The computer reads your frequency and mode and will automatically insert that data into your log as you make contacts.

Radio control is available through Commander allowing direct frequency entry, mode, and filter changes. I prefer using knobs over a mouse, so I minimize this window on my computer screen to save space.

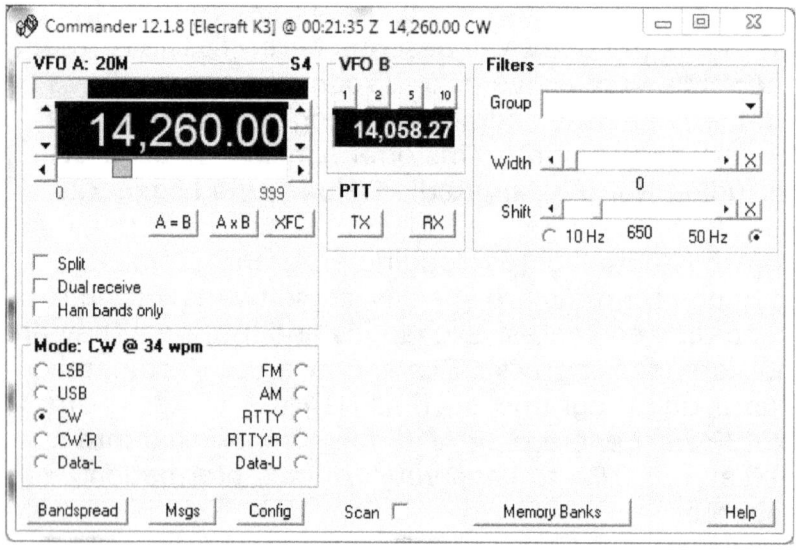

**DxKeeper** displays your logbook. DxKeeper can sort and filter, show the QSL status (sent, received, confirmed, or verified), print QSL labels, and upload your data to LOTW. The Check Progress tab will generate reports showing awards progress.

DxLab is configured with a backup function that automatically saves data to DropBox every time you shut it down. A backup stored in the cloud protects against catastrophes at home. A free DropBox account has more than enough storage.

Here is a screenshot from an incredibly productive couple of days. The contacts with Chesterfield Islands were a tough. I had a hard time hearing them until one evening, there was an opening on 15 meters. I worked both the SSB and CW stations within a few minutes of each other.

The shaded bands are yellow on my screen and indicate the station participates in LOTW. Notice that everyone except the station in Egypt uses LOTW.

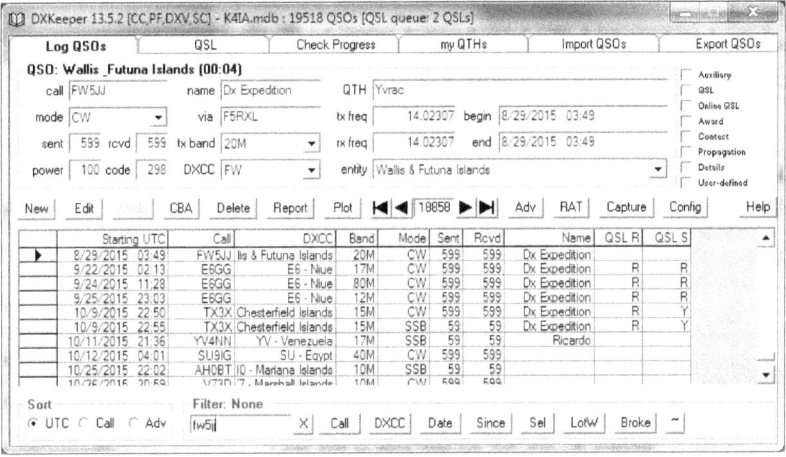

**DxKeeper Capture** is the QSO information entry box. Entering a call in the Capture box, populates other fields with information from the Internet or previous QSOs. When you hit the Log button, the data is put in the logbook portion of DxKeeper, and clears the fields for your next QSO.

The Spot button at the bottom enters the data in the SpotCollector module which distributes to spotting networks for the rest of the world to see. Hint: Don't spot a station until after you work him. Spotting will increase the competition as new stations jump in and try to work him.

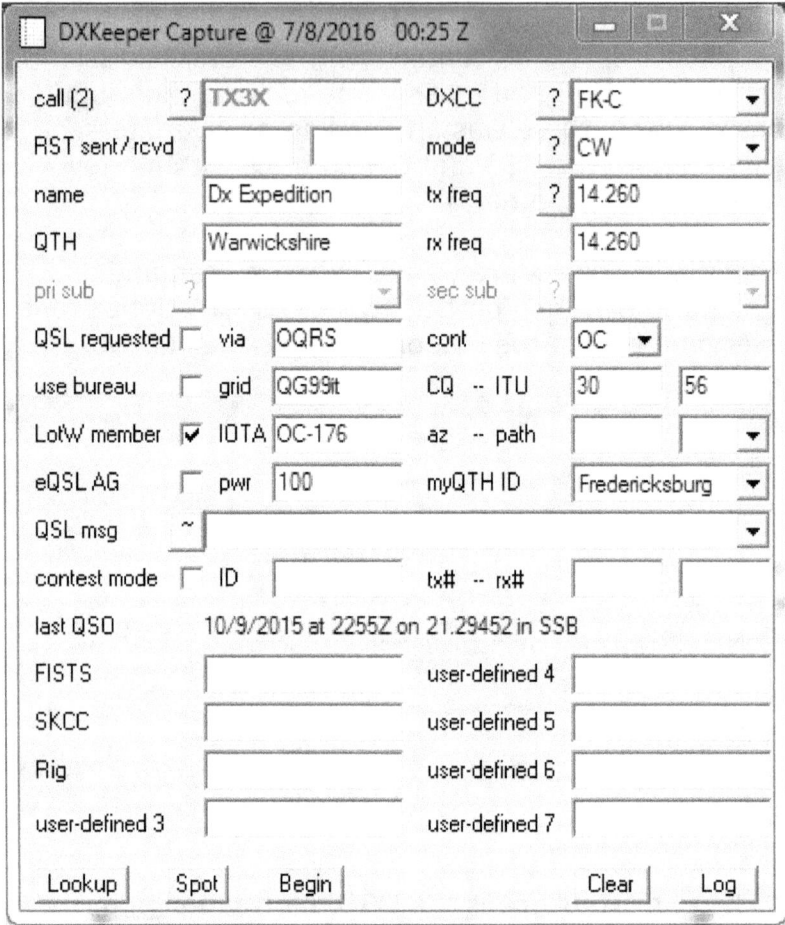

I keep track of changes to my station by entering my callsign in the capture window and putting the changes in the comments section as if I had a QSO with myself. I can search on "K4IA" to see the entries and comments.

**DxView** is a chart based on a DX entity. It shows band and mode progress for that entity, distance, and bearing along with the time zone. There are functions showing the path and local maps. There is also a World Map showing the short and long paths and gray line. You can see the distance to the Chesterfield

Islands (8,887 miles short path) and clicking the ~ symbol gives you the long path distance.

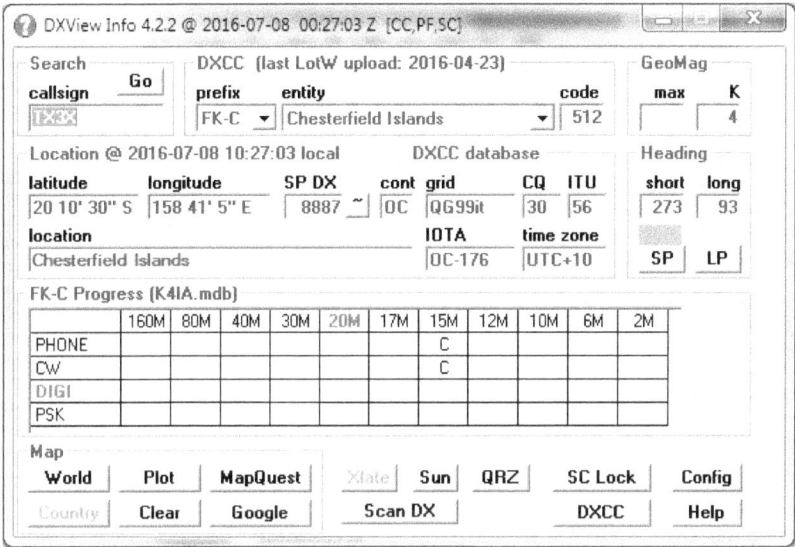

The chart is telling me I have worked TX3, Chesterfield Islands, twice, and have received confirmation (C). I did not have the contact verified with ARRL when I made this screenshot. I submitted the confirmation to ARRL for my 335th entity, and, once accepted, the C changed to V for verified.

**Pathfinder** is a callbook lookup function. Enter a callsign in the Capture box and the callsign is exported to Pathfinder. Pathfinder looks up the station's data on QRZ.com, Buckmaster, and many other databases. Pathfinder imports the operator's name and QTH into the DxKeeper Capture window (subject to your option to override. It also displays the station's address and QSL route.

Pathfinder will also display data the station uploaded, including, sometimes, the operator's picture.

# COMPUTER LOGGING PROGRAMS

This screenshot shows the QRZ.com page for the Palmyra Islands DXpedition.

**SpotCollector** receives and aggregates data from multiple spotting sources. SpotCollecter alerts you to who is on and if they are working split, who heard them, and who uses LOTW. Different color fonts show DX you need.

Up-to-date solar flux and geomagnetic readings are in the upper left corner.

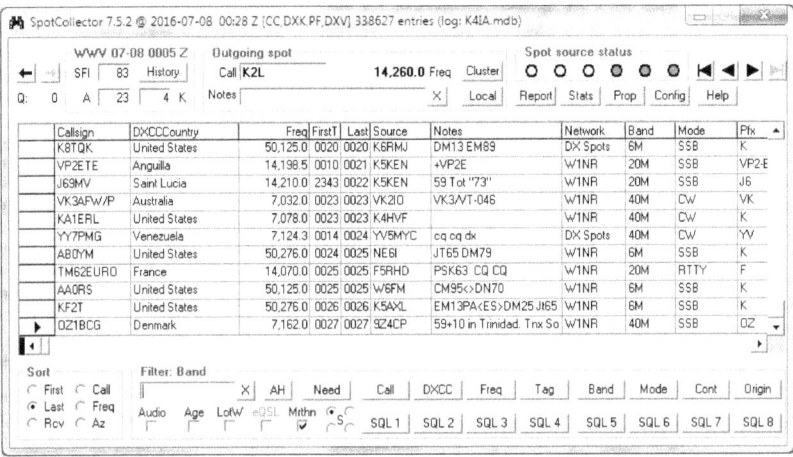

SpotCollector can search for a callsign or entity to see the history of when a particular station has been heard.  You can also choose to see only spots around a frequency or band, mode, entity, or other criteria.

**Winwarbler** is the digital interface supporting sending, receiving and logging PSK, RTTY and other digital modes.  There is a screenshot in the Modes chapter.

**PropView** is a propagation prediction program.  It grabs the current Solar and geomagnetic numbers from  the Internet to generate  a chart.

Here is the prediction for Palmyra & Jarvis Islands.
The dark line at the bottom is the Minimum Useable

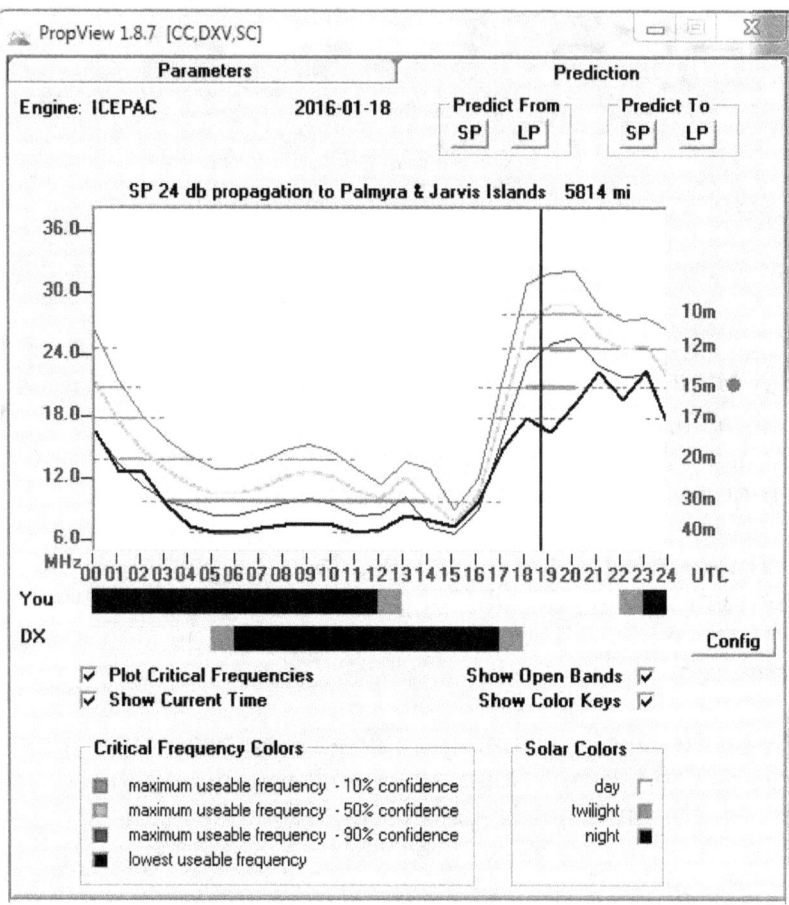

Frequency.  Llighter lines at the top are the Maximum
Useable Frequencies at a 10%, 50%, and 90%
confidence level.  Straight horizontal lines show
predicted band openings, and the vertical line is the
present time.  It looks like 15 meters should be
hopping right now.  Unfortunately, I don't hear
anything.   Predictions are just predictions, not
certainties.

# CONTESTS

Why contest?  Can you imagine a more target-rich environment?  Lots of stations are on the air looking for contacts.  Contesters are eager for every single QSO they can muster.  They will listen and try to dig you out because your QSO means points for them.

We are all competitive.  It's fun to compete against other operators or our last year's score.  Contests reveal what we can do better.

Contests stress-test the station and ourselves.  We don't want to break anything, but it helps to know what works and what fails.  Stretch and grow.  Try a new antenna or operating strategy.

Contest clubs are a way for equipment-challenged hams to band together and create super-stations for their members.  Here's a card from the Saipan BBQ Contest Club.

If the bands aren't cooperating, fire up the grill and barbeque.

Contests will provide an excellent chance to work a lot of stations and entities in a short period. Many contesters upload their logs to LOTW as soon as the contest ends. You get quick confirmations without sending a QSL card.

So, how does a non-contester operate in a contest? First, identify the contest. There are a dozen every weekend. Read the Contest Corral column in *QST* magazine. Eham.net and NG3K.com have lists of upcoming contests with links to the sponsor and rules. Also, check out the WA7BNM contest calendar online. The smaller contests may have suggested operating frequencies.

There are many DX contests, including those sponsored by different countries. Most states have a QSO party during the year. My state of Virginia hosts a popular QSO party in mid-March.

Second, learn the exchange. What information does the contestant want? That will be on the contest website, or you can listen and figure it out. Typically, it is a signal report (always 59) and the contact serial number.  Maybe it includes a CQ Zone Number, state, or power.  In a state QSO party, the typical exchange for someone in the state is a serial number and their county.  An out-of-state station would give a serial number and their state.

Third, get on the air.  You are not trying to win, so don't get discouraged that someone sends you a serial number in the thousands when you are just getting started.  Your QSO #2 is worth just as many points as QSO #1046.

In the end, send the log to the contest sponsor via  an electronic upload from the computer logging program.  No one will laugh at your low score.  You may qualify for a certificate.  I always win the VA QSO Party certificate for the highest score in Fredericksburg because I'm the only one who enters.  I get a nice certificate by default.[37]  (Don't tell anyone).

Here are a few contest operating tips:

Contests are about an exchange.  Nothing more.  Saying "Hi" to an old friend is OK, but no one wants to chit-chat or talk about your weather.

Contest exchanges are crisp and quick.  Here are some examples of things not to say because they are unnecessary, repetitive, and slow down the pace.  I bet you will hear every one of these admonitions violated.  Do it right, and you will sound like a pro.

Don't say "Roger the 59, number 1046 from Ohio." He knows the exchange he gave you.  Ask for a repeat if you didn't get it, but there is no need to parrot it back.

---

[37] I've also won first place in the VA QSO Party for the single-band, single-operator and multi-operator categories.  Those were not by default.

## CONTESTS

Don't say, "Please copy."
Give the exchange; he already knows he has to copy the exchange. "Please copy" is like nails on a chalkboard.

Don't say, "When last heard."
What else would you be reporting? How he sounded before you last heard him?

Don't give his callsign again.
He already knows his callsign, and he knows you are calling him because he acknowledged you.

Don't give your callsign again.
He got it already. If you repeat it, he is going to wonder if you are trying to make a correction.

If you call CQ and have several people come back, don't say, "QRZ?" The same three will call again and you won't be able to sort them out the second time, either. Say QRZ a third time and they will all give up be gone. Concentrate to pick out something distinct. Say, "Who is the K4?" Work him and thin out the herd.

Two contesting strategies are "search and pounce" and "running." Search and pounce means tuning and working stations calling CQ. I've watched Field Day ops twiddling the dial in a trance, afraid to call. It is not enough to search. You must pounce. "Running" is holding a frequency and calling CQ yourself (also called "park and bark"). Running requires grit and determination to fend off those who would try to steal the frequency. Having a big signal helps.

# AWARDS & SPECIAL EVENTS

If you like to collect wallpaper (certificates) and lumber (plaques), you picked the right hobby. There are hundreds of awards available.

A popular award to go for first is WAS, Worked All States, sponsored by ARRL. A worthy endeavor but there are lots more. ARRL also sponsors DXCC (DX Century Club) for confirming contact with 100 DX entities. There's an impressive certificate for confirming all continents (WAC).

Most awards have subcategories for single bands, single modes, and low power (QRP). Confirm 100 entities on each of five bands to earn the 5BDXCC.

*CQ Magazine* sponsors its version of DXCC. They also sponsor a Worked All Zones (WAZ) award for proof of contacts with all 40 CQ Zones. The WPX award recognizes confirmed QSOs with the many callsign prefixes throughout the world.

Various national and international organizations offer awards for contacting member stations. Some require as few as a single contact. *CQ Magazine* has a monthly column describing new awards from around the world. You can also find information at DXAwards.com.

Special-event stations and famous locations are not exactly awards but can net fascinating QSL cards commemorating historical events and sites. You will hear several special-event stations every weekend.

Anniversary of the 1944 Polish uprising.

Barry Goldwater was a US Senator and 1964 Presidential candidate. As a ham, he spent many hours handling phone patch traffic for soldiers in Viet Nam. I never worked Senator Goldwater in real life, but I have the card from this Special Event Station commemorating his service.

This Austrian station celebrated the European Football Association Championship games.

In addition to special events, clubs adopt and operate from famous locations.

Proportionally, Bedford, Virginia, suffered the highest D-Day losses of any US community. Bedford is host to the National D-Day Memorial.

You can talk to the radio room of the *HMS Queen Mary, USS Wisconsin*, and many other vessels.  The *USS Wisconsin* is berthed in Norfolk, Virginia.

"Big Mo" is stationed at Pearl Harbor, Hawaii.  The Japanese surrender ending WWII took place on its decks.

# LEARNING CW

If I told you to learn a foreign language of only 26 words and count to 10, would you think it an impossible task?  Probably not.

Boy Scouts was my first experience with CW. Unfortunately, I learned it all wrong and had to relearn it as a ham.  What went wrong?

In Boy Scouts, we memorized CW as dots and dashes. The first exposure was visual, not audible.  You had a little chart and to look up every letter.  What's wrong with that?  Your brain has to translate what it sees or hears into dots and dashes and then translate those into letters.  It becomes a multi-step process.  It is like learning to translate from English to Spanish to get to French.

I thought I understood it pretty well until someone sent Morse code by flashlight.  It was blinking light to dots and dashes to letters.  It was even worse when sent by waving a flag.  I was lost.

Radio CW is audible.  Learn to recognize the sound and not memorize dots and dashes.  "A" is not dot-dash.  It is not even dit-dah.  It is the sound of dit-dah.   Learning the sound eliminates all the in-between translations.

For the same reason, do not learn that "A" sounds like "Ah-pull" or that the letter "A" has a short line and a long line.  I've seen pictograms like children's alphabet blocks.  These gimmicks introduce additional mental steps.  Now you are going from English to Spanish to French to get to Russian.

Learn CW by hearing it one or two letters at a time until you can immediately make the connection from

## LEARNING CW

the sound to the letter.  Then conquer the next letters and build.  This is called the Koch method.

Another impediment to my learning was the way we sent CW.  At slow speeds "A" was diiiiiiit-daaaaaaaaaah.  Then, the next letter came immediately, with no time for the brain to work.  The modern way is called the Farnsworth timing method.

Farnsworth timing is to send the letters at least 20 words per minute[38], so each letter has one distinct sound.  Dits and dahs may form a letter, but you should not try to hear individual elements.  The letter is one sound.

To slow down the sending pace, Farnsworth increases the space between letters and words.  This does two things.  It reinforces a single sound as the letter, and it gives the brain extra time between letters and words for the translation.  You should also send using Farnsworth timing because that is how the other guy learned as well.  I set my keyer around 22-26 words per minute and slow down by increasing spaces.

After learning the letters, you will recognize words.  When you read, you do not see letters; your brain jumps to the word.  "Word" is not "W-O-R-D."  The same happens with Morse code.  You learn to recognize your callsign, RST, 5NN, TU, 73 and other common "words" without thinking about the individual letters or the elements that make up the letters.

This all takes practice.  In the beginning, listen to a code practice CD or audio file.  K7QO offers a free course download on his website, K7QO.net.  G4FON has a Koch trainer at G4FON.net.  There are other sources, as well.

---

[38] Words per minute (WPM) is based on a 5 letter word. "Paris" is an often-used standard.  Send "Paris" 20 times in a minute for 20 WPM.

Once you get the letters down (mostly), listen to QSOs to learn the standard QSO pattern. Then, GET ON THE AIR! There is no better practice than making actual QSOs. Real QSOs are exciting and won't seem like practice.

Getting on the air is the best way to practice.

FISTS CW Club and Straight Key Century Club (SKCC) promote CW, and you will find slow CW on the frequencies suggested on their web pages. Both FISTS CW Club and SKCC assign you a member number you exchange with other members to collect awards. It is a fun challenge.

Suggested frequencies to find slower CW include:
3.550 – 3.570 MHz
7.055 – 7.060 MHz
14.055 – 14.060 MHz
21.055 – 21.060 MHz
28.055 – 28.060 MHz

# FISTS USA
### FISTS Number 10,000

# KNOWCW

### USA Club Call of FISTS CW Club
### Headquarters: PO BOX 47, Hadley MI 48440

| Radio | Date | UTC | MHz | 2-X | RST | Power | QSL |
|-------|------|-----|-----|-----|-----|-------|-----|
| K4 1A | 5/13/06 | 1758 | 7MHz | cw | 599 | 100 w | Tnx Pse |

FIST members can get permission to use the FISTS CW Club callsign.

When starting out on CW, concentrate on the QSO Trinity first. The pattern is always the same, so you know what to expect. With time, you will progress into more complex conversations. Try to match the speed of the other guy, but if you can't, "QRS" means "slow down" and "QRQ" means "speed up."

For more serious practice, tune to the W1AW Code Practice Sessions. These texts from *QST* magazine are harder to copy because the words are longer and not as predictable. ARRL offers code proficiency certificates that will look impressive on your wall. http://www.arrl.org/w1aw-operating-schedule

Greetings from QRP Station

# K5QLF

Fred Bonavita
334 Royal Oaks Drive
San Antonio, Texas 78209
Bexar County

Ex: W5QJM, W7JLX, W4WUQ, ZF2AL
G-QRP 819; QRP ARCI 4577; Nor Cal 69
CW Operators QRP Club 31; FISTS 3231
SMIRK 3476 (EL09) *CC = 990*

*Power is no
substitute for skill*

To: K4 IA
OP: Buck

When someone has a lousy fist[39], you joke, "QLF? Are you
sending with your left foot?"

I earned this certificate for demonstrating proficiency while
sending with my left foot.

---

[39] "Fist" refers to the operator's style of sending; his "accent" when sending
CW.

For the ultimate challenge, jump into a CW contest.
Contests are almost guaranteed to add 5-10 WPM to your
speed.  I am not sure that Bob had CW in mind when he
made this jump.

YV5DEH in Venezuela is inviting you to get on HF.

# SUMMARY AND A FINAL THOUGHT

This book promised to show you The Easy Way to get on HF. Here is a summary of The Easy Way methods:

- Get involved with a local club.
- Find an Elmer or two.
- Listen, listen and learn.
- Get on the air no matter how modest your station or antenna.
- Don't despise humble beginnings.
- Call CQ after you are comfortable.
- Get on digital modes such as PSK31 and FT8.
- Upgrade your antenna first before upgrading power with a mid-range amplifier.
- Work some contests and state QSO parties.
- Use LOTW.
- Use a computer logging program.
- Be courteous.

Modern amateur radio equipment is far more capable and inexpensive than in the past. This is the Golden Age of ham radio. Get on the air and enjoy it.

73/DX
Buck
K4ia

# APPENDIX A – ANTENNA LENGTHS

Values for each half of a dipole based on 234/frequency = length in feet.  Cut longer and adjust.  To adjust, note the actual and target frequency of lowest SWR.  Add or subtract the difference in length for each half of your antenna.  To move lower, add wire.  To move higher, wrap it back. Bold frequencies are the approximate bands.

| MHz | Feet | Inches | MHZ | Feet | Inches |
|-----|------|--------|-----|------|--------|
| 3.2 | 73.2 | 878 | 10 | 23.4 | 281 |
| 3.3 | 71 | 851 | 10.05 | 23.3 | 279 |
| 3.4 | 68 5/6 | 826 | **10.1** | 23.2 | 278 |
| **3.5** | 66 6/7 | 802 | **10.15** | 23 | 277 |
| **3.6** | 65 | 780 | 10.2 | 23 | 275 |
| **3.7** | 63.25 | 759 | 10.25 | 22.8 | 274 |
| **3.8** | 61.6 | 739 | 10.3 | 22.7 | 273 |
| **3.9** | 60 | 720 | 10.35 | 22.6 | 271 |
| **4.0** | 58.5 | 702 | 10.4 | 22.5 | 270 |
| 4.1 | 57 | 685 | 10.45 | 22.4 | 269 |
| 4.2 | 55 .7 | 669 | 10.5 | 22.3 | 267 |
|  |  |  |  |  |  |
| 6.7 | 35 | 419 | 13.9 | 15.8 | 202 |
| 6.8 | 34.4 | 413 | **14.0** | 16.7 | 201 |
| 6.9 | 34 | 407 | **14.1** | 16.6 | 199 |
| **7.0** | 33 3/7 | 401 | **14.2** | 16.5 | 198 |
| **7.1** | 33 | 395 | **14.3** | 16.3 | 196 |
| **7.2** | 32.5 | 390 | **14.4** | 16.25 | 195 |
| **7.3** | 32 | 385 | 14.5 | 16.1 | 194 |
| 7.4 | 31 5/8 | 379 | 14.6 | 16 | 192 |
| 7.5 | 31.2 | 374 | 14.7 | 16 | 191 |
| 7.6 | 30.8 | 369 | 14.8 | 15.8 | 190 |
| 7.7 | 30.4 | 365 | 14.9 | 15.7 | 188 |

| MHz | Feet | Inches | MHZ | Feet | Inches |
|---|---|---|---|---|---|
| 17 | 13.75 | 165 | 24 | 9.75 | 117 |
| 17.5 | 13.4 | 160 | 24.8 | 9.4 | 113 |
| 18 | 13 | 156 | **24.89** | 9.4 | 113 |
| **18.068** | 13 | 155 | **24.94** | 9.4 | 113 |
| **18.168** | 12.8 | 155 | 25.1 | 9.3 | 112 |
| 18.25 | 12.8 | 154 | 25.2 | 9.3 | 111 |
| 18.35 | 12.75 | 153 | 25.3 | 9.25 | 111 |
| 18.45 | 12.6 | 152 | | | |
| 18.55 | 12.6 | 151 | | | |
| 18.65 | 12.6 | 151 | | | |
| 18.75 | 12.5 | 150 | | | |
| | | | | | |
| 20.8 | 11.25 | 135 | 27.5 | 8.5 | 102 |
| 20.9 | 11.2 | 134 | **28** | 8.3 | 100 |
| 21.0 | 11.1 | 134 | **28.2** | 8.3 | 100 |
| **21.1** | 11 | 133 | **28.4** | 8.25 | 99 |
| **21.2** | 11 | 132 | **28.6** | 8.2 | 98 |
| **21.3** | 11 | 132 | **28.8** | 8.1 | 98 |
| **21.45** | 11 | 131 | **29** | 8 | 97 |
| 21.55 | 10.8 | 130 | **29.2** | 8 | 96 |
| 21.65 | 10.8 | 130 | **29.4** | 8 | 96 |
| 21.75 | 10.75 | 129 | **29.6** | 8 | 96 |
| 21.85 | 10.7 | 129 | 30 | 7.8 | 94 |

# APPENDIX B - ABBREVIATIONS AND PROSIGNS* FOR CW

| ABT | about |
|---|---|
| AGN | again |
| ANT | antenna |
| AR* | end of message |
| B4 | before |
| BCNU | be seeing you |
| BK* | break (sent as one letter) |
| BTR | better |
| BURO | bureau (QSL buro) |
| C | yes |
| CFM | confirm |
| CK | check |
| CPY | copy |
| CUD | could |
| CUAGN | see you again |
| CUL | see you later |
| DE | this is |
| DR | dear |
| ENUF | enough |
| ES | and |
| FB | fine business |
| GB | good bye |
| GM, GA, GE, GN | good morning etc. |
| GG | going |
| GUD | good |
| HI HI | ☺ LOL laughter |
| HR | here or hear |
| HV | have |
| HW | how |
| KN* | no one else answer |
| LID | poor operator |
| LIL | little |
| N | nine (RST 5NN) |
| OM, OT, OC, OB | old man, top, chap, boy |
| OP | operator or name |
| PSE | please |
| PWR | power |

| R | roger, OK |
|---|---|
| RCVR, RX | receiver |
| RIG | equipment |
| RPT | repeat or report |
| SASE | self-addressed stamped envelope |
| SED | said |
| SHD | should |
| SIG | signal |
| SK* | end of conversation |
| SKED | schedule |
| T | zero (pwr 1TT w) |
| TFC | traffic |
| TMW | tomorrow |
| TT | that |
| TNX,TKS | thanks |
| TU | thank you |
| TX, XMTR | transmitter |
| U, UR, URS | you, your, yours |
| WID | with |
| WK, WKD | work, worked |
| WL | will |
| W,WTS | watts |
| WUD | would |
| WX | weather |
| XCVR | transceiver |
| XTAL | crystal |
| XYL | wife |
| YL | young lady |
| 73 | best regards |
| 88 | love & kisses |

*ProSigns are sent as one letter

# APPENDIX C - Q SIGNALS

Q signals become a question when followed by a question-mark. QTH? Where are you located?
Answer: QTH VA.

QRG   Frequency.

QRL   Is the frequency in use?

QRM   I am troubled by interference.

QRN   There is noise / static.

QRO   Increase power.

QRP   Decrease power or low power.

QRQ   Send faster.

QRS   Send slower.

QRT   Stop sending or I am going off the air.

QRU   I have nothing further for you.

QRV   I am ready.

QRX   Standby.  QRX 10.  Standby 10 minutes.

QRZ   Who is calling me?

QSB   Signals are fading up and down.

QSK   I can hear you between my signals.  You can break in during my transmission.

QSL   Acknowledge receipt.

QSO   Contact.

QSY   Change to another frequency.

QSX   Listening frequency

QTH   Location.

# INDEX

# INDEX

www.ingramcontent.com/pod-product-compliance
Lightning Source LLC
Chambersburg PA
CBHW070245190526
45169CB00001B/305